INTERNATIONAL SERIES OF MONOGRAPHS IN
NATURAL PHILOSOPHY

GENERAL EDITOR: D. TER HAAR

VOLUME 21

CORRELATIONS AND ENTROPY
IN CLASSICAL STATISTICAL MECHANICS

OTHER TITLES IN THE SERIES
IN NATURAL PHILOSOPHY

CORRELATIONS AND ENTROPY IN CLASSICAL STATISTICAL MECHANICS

BY

J. YVON

TRANSLATED BY

H. S. H. MASSEY

EDITED BY

D. TER HAAR

THE QUEEN'S AWARD
TO INDUSTRY 1966

PERGAMON PRESS

OXFORD · LONDON · EDINBURGH · NEW YORK
TORONTO · SYDNEY · PARIS · BRAUNSCHWEIG

Pergamon Press Ltd., Headington Hill Hall, Oxford
4 & 5 Fitzroy Square, London W.1

Pergamon Press (Scotland) Ltd., 2 & 3 Teviot Place, Edinburgh 1

Pergamon Press Inc., Maxwell House, Fairview Park, Elmsford, New York 10523

Pergamon of Canada Ltd., 207 Queen's Quay West, Toronto 1

Pergamon Press (Aust.) Pty. Ltd., 19a Boundary Street, Rushcutters Bay, N.S.W. 2011, Australia

Pergamon Press S.A.R.L., 24 rue des Écoles, Paris 5^e

Vieweg & Sohn GmbH, Burgplatz 1, Braunschweig

First English edition 1969

This is a translation of the original French
Les Corrélations et l'Entropie en Mécanique Statistique Classique
published by Dunod, Paris, 1966

Library of Congress Catalog Card No. 68-25570

PRINTED IN GERMANY

08 012755 ×

Contents

Contents

Introduction

A MASS of water is made up of a large number of molecules. This structure, which is not revealed in any way by a casual glance, is nevertheless responsible for this liquid's more familiar properties of fluidity, expansion, vaporization, surface tension and transparency. It governs both its static and its dynamic behaviour: various forms of laminar or turbulent flow, heat transmission, acoustic properties. The molecular theory of water attempts to explain, as a function of its structure, the properties displayed by this fluid when it is observed in noticeable amounts on the human scale, i.e. in macroscopic amounts. The same problem arises for all specimens of matter, whatever their state may be.

The word molecule must be understood here in its broad sense: it is more generally a question of the ultimate elements of matter that the theoretician is willing to introduce explicitly in his arguments. It is in this way that the theory of the compressibility of gaseous nitrogen was established by looking upon the gas as being made up of indestructible molecules of nitrogen. The examination of other phenomena, for example the passage of electric charges through a gas, makes us call on smaller elements; we have to imagine, as well as the nitrogen molecules, the ions, the electrons, the free atoms coming from their dissociation. When interpreting any macroscopic phenomenon, therefore, we must make a choice regarding the particles to be considered as the elementary constituents of the matter being studied.

Mechanical systems are systems made up of very numerous particles. Provided that the forces exerted on these particles and the forces that they exert on each other are known, the equations of mechanics make it possible to calculate the future behaviour of the system when the initial state has been fixed. A problem of this kind is in fact insoluble because of the enormously high number of parameters concerned. Only general theorems can be stated on this subject. In actual fact, however, physicists do not have to have a precise knowledge of the movement of a system starting at a certain initial state. The experimenter cannot, indeed, determine the state of the system, whether it be the initial state or subsequent states. The number of quantities that he can measure is infinitely smaller than the number of parameters necessary for describing the fate of each particle. It is a question here of problems in mechanics where there is a certain uncertainty about the observed state. As a result, to make the connexion between the small-scale phenomena and macroscopic observation, probability considerations and statistical considerations have to be superimposed on mechanical considerations.

The correct principles of mechanics (those which conform to physical reality as it is known at present) are given by quantum mechanics. The state of a quantum system, in the sense of the preceding paragraph, is described by its wave function. It must be admitted that the wave function of a macroscopic system is rather badly determined. Therefore the treatment of the problems of interest to us here involves a double uncertainty: the one that is fundamental to quantum mechanics and the additional uncertainty that we express when saying that the wave function is badly defined; the combination of these two effects is taken into consideration by the quantum theory of mixtures: the latter are the basis of developments in quantum statistical mechanics. This superimposition of uncertainties means that mechanical quantities can be observed only as mean values.

In certain cases, however, classical mechanics (that is, prequantum mechanics) are a sufficient approximation of quantum mechanics. The state of a classical system is defined by stating the sets of spatial and momentum coordinates of the system. In the molecular theory of macroscopic systems a state of this kind is never well defined, there being only a probability law which assigns a large or small weight to the different possible states. This does not mean that the laws of mechanics are modified in any way. The system describes the trajectory on which it is initially started in the usual way, but we do not know exactly which trajectory it is on. In other words, the mechanics with which we are dealing are definitely classical mechanics with the restriction that they are subject to uncertainty about the initial conditions. These problems are treated by classical statistical mechanics. When there is uncertainty about the initial conditions the notion of a trajectory loses much of its interest and disappears from applications as soon as the system in question is complex. Classical statistical mechanics, just like quantum theory, lead also to an estimate of averages. Classical statistical mechanics are the approximation of the theory of quantum mixtures when Planck's constant is negligible. Because of its simplicity we shall first use this approximate treatment in our undertaking of interpreting molecularly the properties of macroscopic media.

Classical statistical mechanics and the quantum theory of mixtures (also called quantum statistical mechanics) are mechanical treatments which can be studied independently of their application to the molecular theory of macroscopic media. We shall start, moreover, by building up the formalism of the one or the other without worrying at first about applications to macroscopic systems. In this phase of the study there is no question of being preoccupied with the complexity of the mechanical system: it is a question of theories which can even be applied to a single particle subject to a field of force.

As soon as we look at macroscopic physical systems the immediately dominating fact is that the simplest macroscopic systems contain molecules of only a single kind. The mathematical formalism which we shall be using assigns a number to the molecules, and this seems to give them a certain individuality which experience does not confirm. Molecules of the same kind are indistinguishable. There is, in fact, at a certain point in the theory a compensating factor which completely eliminates the effects of this unjustified numbering. This said, the mechanical averages which have to be considered are all the same in nature: they are sums of equivalent terms, each of which concerns only one molecule at a time, or two molecules at a time and rarely more. The result is that the evaluation of these averages does not require the knowledge of the fundamental distribution functions, the classical phase density or the quantum density operator, which are generally highly complicated, but of simpler functions of which it is at times possible to form an idea without having recourse to any complex theory: the molecular density, or, if we cut out any attempts at pretentious language, the density is a typical example of this set of distribution functions with modest but effective pretensions. The procedure which allows us to pass from the fundamental distribution functions, which depend on an immense number of parameters and cannot be used directly, to the simple distribution functions (also called reduced quantities) is called the regressive procedure. The situation is not seriously altered when the system, instead of containing a single kind of particle, contains only a limited number of kinds.

When applied within the framework of quantum theory the regressive method has a different interest: it allows us to show that consideration of the case when the wave function is badly defined does not add anything more to this theory. There is nothing fundamental in the superposition of uncertainties which it has suited us to call on above: all the uncertainties which we might be concerned with in quantum theory are a consequence of the general principles of this doctrine. There is no need, therefore, to think that the requirements of the molecular theory of macroscopic media lead to adding supplementary elements on to quantum theory. Thus quantum mechanics "contains" statistical mechanics. To put it briefly, we could say that quantum statistical mechanics does not exist: it has no true autonomy. Nevertheless to leave it a hint of individuality let us say that quantum statistical mechanics is a method of studying systems which contain a large number of particles. This subservience of statistical mechanics is less evident if we keep to the classical point of view alone: but there is no point in insisting on this since the classical point of view is not fundamental.

Ordinary matter often appears to the observer in a state of rest. This state of rest, carefully defined by the physicist, is the state of thermodynamic equilibrium. In fact molecular theory interprets this apparent rest by an underlying active movement of the constituent particles. This is the phenomenon of thermal motion. The rest concerns only the macroscopic averages which result from statistical evaluations: they are time-independent. The theory of thermodynamic equilibrium is one of the principal successes of molecular theory. The idea of temperature in particular has its interpretation in the framework of the principles of mechanics and of the hypotheses of statistics.

To describe equilibrium we must first examine the case when the system's Hamiltonian is time-independent. In addition, among the uncertain factors we must assume that the energy is not well defined. This said, the theory of equilibrium rests essentially on the idea that the probability of a microscopic state is a function of the energy. The fact that the law of probability involves only energy means, as is evidently indispensable, that the corresponding macroscopic state is stationary.

An isolated thermodynamic system which is not in equilibrium tends towards thermodynamic equilibrium. At least, this is what experience teaches us. The interpretation of this phenomenon presents great difficulties, some mathematical and some physical and so intimately mixed it is difficult to recognize them. The case of gases, that is the one where the evolution is dominated by binary collisions of molecules, was treated long ago by Boltzmann (1896). His treatment is based on the hypothesis of molecular chaos. It leads to an explanation which is quantitatively as perfect as the calculations allow for the phenomena of viscosity and thermal conductivity. The molecular chaos hypothesis had aspects which can be criticized, but these criticisms have lost their weight today. The problem is to reconcile the reversible nature of the equations of mechanics—it is a question of reversibility in relation to time—with the irreversibility of the natural phenomena: much effort has been spent in this direction to "fool" mechanics, but it appears that the solution of the paradox is elsewhere: one is led to admit that the initial state of the system is not any particular one, but that it is a state of great disorder. This is a statistical hypothesis which in no way contradicts the laws of mechanics. Analysis of the implications of this hypothesis is far from being complete at present—the fact that it is probably satisfactory does not mean that it is sufficient—but it is certainly a strong guide line.

The study of irreversible phenomena necessitates a formalism which can be explained more economically in quantum theory. That is why this study will be incorporated only in a volume which I hope to devote to this theory. Anyhow, in the present volume we shall be touching on the problem of an essential quantity in this matter—statistical entropy—and we shall start by examining its properties, which are sometimes those of the entropy of thermal researchers and which at other times differ from them considerably. This will be the opportunity of re-examining the celebrated H-theorem of Boltzmann resulting from the molecular chaos hypothesis: perhaps this will be the opportunity of reconciling certain opponents to the latter.

Introduction

Another type of uncertainty should be added to those which we have considered up to now. This new uncertainty relates to the number of molecules present in the system. Contrary to what one might think at first, this additional uncertainty does not complicate the problem; on the contrary it eliminates small difficulties which may be secondary but are irritating. It is this uncertainty which makes it particularly certain that there cannot often be any correlation between distant molecules: in the other case, even for long distances there will be small correlations and one cannot state that they are always negligible. Above all, it is a question only of matters of convenience and it should be added that the uncertainty in the number of particles is clearly fundamental when the latter are not indestructible, when we admit that they can be transformed into each other.

Let us stress here what "experience" teaches the theoretician: he can only gain by introducing into the concepts which guide him all the uncertainties that the real experimenter cannot master. The problems are so much simplified by this that it is perhaps permissible to parody a celebrated principle by writing the following inequality:

uncertainties tolerated by the theoretician ×
   nuisance he is subject to > universal constant.

The quantities relating to a thermodynamic system display a certainty which is rather surprising, given their statistical origin. For example, the pressure of a gas enclosed in a container which is sufficiently resistant and placed in a thermostat is a quantity which can be measured with great precision. It is here that the "law of large numbers" plays a part. It implies that the fluctuations of a quantity such as pressure, of which we measure only the average, are very weak. It is this which makes certain phenomena of normal mechanics (elasticity of a spring, laminar flow of a fluid) display a very striking deterministic character despite the notoriously statistical nature of the behaviour of the constituent molecules. The law of large numbers creates an apparent order from a considerable internal disorder. We shall see that the smaller the number of kinds of constituent molecules in the medium in question the more favourable will be the part played by the large numbers effect. A medium of very complex chemical composition will never display such good definition of the macro-scopic properties as a medium of simple composition.

The study of gases has preceded that of the other states of matter. In these media the movements of the molecules are freer than in condensed states, and the kinetic energy is greater than the potential energy. That is why the molecular theory of gases has for a long time been called kinetic theory, whilst the general theory, which remained vaguer because of the difficulty of the problems it encountered, received the name which we are using here, namely that of statistical mechanics. The kinetic theory of gases allowed methods which were proper to it and it was possible to have the impression of two treatments which were a little strange to each other. The later unification of the points of view resulting from a progressive penetration of kinetic theory by the methods of the other discipline which improved it has made the adjective kinetic a title out of date. The interpretation, at first paradoxical, of the states of rest or ordered motion of gases by thermal motion of the molecules gave the title of "kinetic theory" a freshness that the title of "statistical mechanics" does not have, although it is more all-embracing. Nevertheless it is natural to continue speaking of kinetic theory every time molecules are to be followed individually: such is the case when it is a question of studying the Brownian movement of a colloidal mycella or studying the ioni-zation of the matter along the trajectory of a fast ion. The necessity of limiting ourselves will leave these problems outside our pre-occupations, although it is the quantitative study of colloidal suspensions and of Brownian motion (Jean Perrin, 1909) that made tangible the well-founded nature of the description of "rest" by motion.

Dislocations in solids and the individual observation of atoms in electron microscopes also correspond to a microscopic obser-vation whose interpretation eludes statistical mechanics as we understand it.

The even broader concepts of stochastic theory go beyond statistical mechanics. This theory avoids dealing with any parti-cular mechanics: it is therefore much used when the elementary laws are unknown or when they are too difficult to take into con-sideration. Here we must "stick to" mechanics, the more so because our attitude will be less general and more precise. It should be recalled, nevertheless, that the Onsager relations (see

Introduction

de Groot and Mazur, 1962), which concern irreversible phenomena, have been the subject of extensive theoretical developments: the latter belong more to stochastic theory than to real statistical mechanics.

We note further that stellar dynamics, the study of star systems, originate in considerations which are not so very different from those of kinetic gas theory (see Chandrasekhar, 1960; Pecker and Schatzman, 1959). A certain formalism is common to both fields. Their connexion with experiments nevertheless differs considerably, so much so that they cannot be treated together. Their deductive nature should not make us forget that they are nevertheless slices of physics and that the latter imposes in each case hypotheses which are foreign to a common mathematical structure.

Conceived as a branch of theoretical physics, statistical mechanics permits approximation methods. These approximations relate in particular to the correlations which, depending on the case, are conveniently neglected, as in the case of the molecular chaos hypothesis, or on other occasions must be treated rigorously. The approximations make the treatment of the problems easier. The corresponding methods supplement other methods of approximation which have been long in use, sometimes simply by codifying them better and sometimes by complete innovations. There is a whole body of doctrine on this. It is much more elaborate in classical theory than in quantum theory. This is one of the reasons for not neglecting the study of the former. Another approximation relates to the molecular model.

In the study of statistical mechanics priority must be given today to systems of material points: it is used to treat molecules in the form of rudimentary models. They are rudimentary to the point that there is no question of starting to describe the specific heat of polyatomic gases or of their chemical reactions. As regards the present volume this is not, properly speaking, a restriction since it will be necessary to have recourse to quantum considerations. We know that the quantum theory of gases is satisfactory in this respect. We shall also be leaving aside problems where inconvenient statistics and geometry interact: for example, the form of linear macromolecules as a function of the temperature, or the interactions between permanently electrically polarized molecules such as those of water. A certain amount of realism will be lost. In compensation, by limiting ourselves to particles of an ascetic form we shall succeed in building up fairly completely structures made up of molecules or microscopic particles which will display correctly the macroscopic properties attributed to them by thermodynamics in particular. A simplified molecular model suffices for touching usefully on the major questions of pressure, heat, entropy and temperature. Without making concessions to the complex effect of the interactions we shall show how the forces of cohesion on the one hand and repulsive forces and thermal motion on the other contradict each other and balance out. Priority will be given to fluids over solids because in the latter case the existence of an ordered lattice channels the thermal motion and simplifies its description, although quantization changes its laws. We could nevertheless touch on certain problems where one imagines that the particles are moving on an imposed lattice. Strongly ionized fluids (plasmas) are a particularly satisfactory field of application because of the central forces which reign there. This is a subject in the throes of development.

Radiation will be a major absentee: its classical and quantum aspects form a slightly disparate ensemble the synthesis of which, which must proceed from the quantum point of view, will not be given here.

The present volume is above all a description of a method and of general results. The reader will often be referred to other works for applications and also in respect of controversial questions, which he can in any case ignore on the first reading. I have chosen to discuss here the ideas which seem to me the most satisfactory: but the idea must be driven home that no problem in physics is ever completely solved and that one must always bear in mind contradictory opinions if one wishes to reach a personal opinion.

Of course, knowledge evolves. Ideas become clearer, more general, more complete. Phenomena are observed with greater precision. Media which originally were not available to experimenters, such as highly ionized gases, offer theoreticians a new, promising field of study and give them new worries. However, only exceptionally do we find new means of attacking problems. Such a new means is provided by the modern processes of numerical calculus. These allow us to exploit the algebraic or analytic approximations of theorists. True, this advantage is common to all theories, but I should like to emphasize that huge numerical calculations now are competitive in statistical mechanics. Dynamical kinds of calculations predict, for instance, how a system of a few thousand particles which all can be followed individually will evolve starting from a given, random, initial state. Experience shows that such a system contains enough particles to give a honest picture of a real fluid, provided one restricts one's ambition to describing simple phenomena. Numerical calculations are a kind of experiment, performed at one's desk. This experiment gives results which are more detailed than those from a real experiment. The material investigated has, of course, schematic properties, but they are known exactly. It is possible that unforeseen phenomena, or phenomena more pronounced than those occurring in nature, can show up. In looking for better models this new technique may beat the traditional theorist, who works with compact analytical descriptions, in the search for approximation methods. On the other hand, as a consolation prize, the theorist may find in these results from calculations an efficient means to control his ideas. It seemed useful to me to point out in the present Introduction this new feature which hardly appears in the main body of the present book, but which may become extremely valuable.

Before being published a large part of the material in this volume had been used for some years as teaching material for the students of the Third Cycle of the Paris–Orsay Faculty of Science. I must here thank my colleagues who have allowed me to integrate myself again, after quite a long absence, in the scientific activity of higher education. We know that teaching, whilst it is definitely fruitful for the students, is also good for the teachers whose thinking is thus subjected to an incomparable refining: the students who have followed these courses must receive the ultimate thanks.

Note on Notation

POSITION vectors are indicated by r, their components by x, y, z.

Velocity vectors are indicated by c, their components by u, v, w.

Momentum vectors are indicated by p, their components by p_x, p_y, p_z or by p, q, r.

Forces are indicated by F, their components by X, Y, Z.

Summation over repeated indices is understood to be implied, but this is often written in a symbolical form:

$$\alpha_y p_{xy} \equiv \sum_{\lambda = x, y, z} \alpha_\lambda p_{x\lambda}$$

$$x \frac{\partial f}{\partial x} \equiv x \frac{\partial f}{\partial x} + y \frac{\partial f}{\partial y} + z \frac{\partial f}{\partial z}$$

$$(x_i - x_j)(y_i - y_j) \frac{\partial^2 f}{\partial x \, \partial y}$$

$$\equiv (x_i - x_j)^2 \frac{\partial^2 f}{\partial x^2} + 2(x_i - x_j)(y_i - y_j) \frac{\partial^2 f}{\partial x \, \partial y}$$

$$+ 2(x_i - x_j)(z_i - z_j) \frac{\partial^2 f}{\partial x \, \partial z} + (y_i - y_j)^2 \frac{\partial^2 f}{\partial y^2}$$

$$+ 2(y_i - y_j)(z_i - z_j) \frac{\partial^2 f}{\partial y \, \partial z} + (z_i - z_j)^2 \frac{\partial^2 f}{\partial z^2}$$

$$\frac{p}{m} \frac{\partial}{\partial x} \equiv \frac{p}{m} \frac{\partial}{\partial x} + \frac{q}{m} \frac{\partial}{\partial y} + \frac{r}{m} \frac{\partial}{\partial z}$$

$$X \frac{\partial}{\partial p} \equiv X \frac{\partial}{\partial p} + Y \frac{\partial}{\partial q} + Z \frac{\partial}{\partial r}$$

$$\frac{\partial}{\partial x} n(u) \equiv \frac{\partial}{\partial x} n(u) + \frac{\partial}{\partial y} n(v) + \frac{\partial}{\partial z} n(w)$$

Note on Notation

$$\frac{\partial}{\partial y} pq \equiv \frac{\partial}{\partial x} pp + \frac{\partial}{\partial y} pq + \frac{\partial}{\partial z} pr$$

$$\frac{\partial}{\partial y} uv \equiv \frac{\partial}{\partial x} uu + \frac{\partial}{\partial y} uv + \frac{\partial}{\partial z} uw$$

$$\frac{\partial}{\partial y} k_{xy} \equiv \frac{\partial}{\partial x} k_{xx} + \frac{\partial}{\partial y} k_{xy} + \frac{\partial}{\partial z} k_{xz}$$

$$\frac{\partial}{\partial y} j_x v \equiv \frac{\partial}{\partial x} j_x u + \frac{\partial}{\partial y} j_x v + \frac{\partial}{\partial z} j_x w$$

$$\frac{\partial Y}{\partial y} = \frac{\partial X}{\partial x} \equiv \frac{\partial X}{\partial x} + \frac{\partial Y}{\partial y} + \frac{\partial Z}{\partial z}$$

CHAPTER 1

Probability Densities and their Evolution

1.1. Probability Densities and Averages

We shall first consider a system made up of indestructible particles. The mechanics which govern the movements of the particles are classical mechanics. At a certain given time the position of each particle is fixed by a certain number of parameters. In the simplest case the particle is a material point; then these parameters are the coordinates and the components of the velocity or, better still, of the momentum. Even when the particle is complex we are often content to describe its position by that of its centre of mass. Generalization to more complicated cases is not without difficulty but does not introduce any essentially new ideas from the statistical point of view, provided we keep to the classical point of view. Therefore we shall not be stressing this kind of problem here.

The number of particles in question is N. To make the language more realistic we shall assume that these particles are molecules. Later we shall be assuming that the number N is very high, but for the moment we shall not make any hypothesis in this respect. The molecules are numbered from 1 to N.

When, at a point in time, the coordinates and the components of the momentum of all the molecules are given we say that the system is in a well-defined phase. The phase can be represented in a $6N$-dimensional space. The volume element $d\Omega$ of this space is called the element of extension in phase. Excluding all information on the momenta, the set of the molecules' positions defines the configuration (representation in $3N$ dimensions, extension $d\Omega_c$). Likewise the set of the momenta defines a point in the complementary space (element $d\Omega_p$).

1

Probability Densities and their Evolution

In the simplest cases the momentum is proportional to the velocity: $p = mu$.[†]

The experimenter would not know how to take an interest in a particular phase. For various reasons he measures only *averages*. These averages can be calculated by means of the density of probability in phase:

$$D(r_1, r_2, ..., r_N; \; p_1, p_2, ..., p_N, t).$$

Given its probability significance, the density in phase may be zero but it is never negative.

We write:

$$d\Omega_c = d^3r_1 \, d^3r_2 \, ... \, d^3r_N, \quad d\Omega_p = d^3p_1 \, d^3p_2 \, ... \, d^3p_N,$$
$$d\Omega = d\Omega_c \, d\Omega_p.$$

Then

$$D \, d\Omega$$

is the probability of finding at time t the system in the element of extension in phase $d\Omega$.[††] The density in phase is normalized by the condition

$$\int D \, d\Omega = 1. \tag{1}$$

Condition (1) says that the sum of the probabilities equals unity. In this integration the velocities vary between $-\infty$ and $+\infty$ and the centres of the molecules describe the whole of the volume that is open to them. If we are interested only in the configuration then to calculate the averages it is sufficient to know the density of the probability in configuration

$$D_c(r_1, r_2, ..., r_N),$$

which can be calculated from the integral

$$D_c = \int D \, d\Omega_p. \tag{2}$$

It satisfies the condition

$$\int D_c \, d\Omega_c = 1. \tag{3}$$

The range of the integration is either limited by the walls of the container holding the particles or stretches to infinity. But it is al-

---

† We shall often denote a vector by its first component: x will represent by itself all the coordinates x, y, z, Likewise p represents p, q, r; u represents u, v, w.

†† In the eartly parts of this book, the time does not enter into the argument; we have therefore not always indicated it explicitly.

2

ways assumed that there are very few molecules in remote regions, so that the integral (3) always has a meaning.

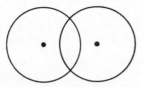

FIG. 1.1. Situation that is impossible in the common sense of the word.

It may happen that the molecules can be treated as hard spheres that cannot be penetrated by one another: in statistical calculations we do not consider a situation like that shown in the figure to be impossible but to have a probability like any other situation. This probability is merely nil. Therefore the surfaces of the molecules are not integration limits. This attitude simplifies enormously the problems of kinetic theory.

Any physical quantity, for example the energy, is represented by a phase function

$$F(r_1, r_2, ..., r_N; \quad p_1, p_2, ..., p_N).$$

We shall always assume that such functions are defined without ambiguity whatever the number of particles present. It can also be explicitly time-dependent. The experimenter will be interested in the average (this modest phrase conceals a monumental problem—that of the theory of measurement)

$$\langle F \rangle_N = \int FD \, d\Omega. \tag{4}$$

When there is no ambiguity the suffix N will be omitted.

According to the case, D will or will not be time-dependent. The same will be true of the averages. It turns out sometimes that we have to evaluate a whole series of averages relating to the same quantity F which are of the form $f(F)$. In this case we are interested in calculating once and for all

$$D(\mathscr{F}) = \int \delta(\mathscr{F} - F) D \, d\Omega, \tag{5}$$

δ being the Dirac δ-function here. From this we have

$$\langle f(F) \rangle = \int D(\mathscr{F}) f(\mathscr{F}) \, d\mathscr{F}. \tag{6}$$

3

It is often quite difficult to use this formula in a practical manner. Let us put

$$\Delta F = F - \langle F \rangle,$$

i.e. ΔF is the fluctuation of F. The average is obviously zero. We shall often be content to calculate the variance of F, namely

$$\langle \Delta F^2 \rangle = \langle (F - \langle F \rangle)^2 \rangle = \langle F^2 \rangle - \langle F \rangle^2. \tag{7}$$

A case that is important and with which we shall now deal is that where all the particles are identical. They are indistinguishable and the act of numbering them has no physical significance. We shall not give up the numbering but by imposing symmetry properties on the density in phase eliminate any possible errors: it is invariant when the parameters relative to any two molecules are permuted.

The averages that concern especially the configuration deserve a first examination.

1.2. Molecular Density

Let us consider in the volume V open to the molecules a fraction A of this volume.

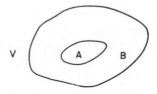

Let us define the function F_J relating to the molecule J and equal to 1 when the centre of the molecule is in A, but equal to zero when its centre is in B. For a given configuration the number of molecules contained in A is

$$N_A = \sum_1^N F_J.$$

The average of this number is

$$\langle N_A \rangle = \int \left(\sum F_J \right) D \, \mathrm{d}\Omega$$

or, what is the same,

$$\langle N_A \rangle = \sum \int F_J \, D_c \, \mathrm{d}\Omega_c.$$

As all the molecules are equivalent this is also

$$\langle N_A \rangle = N \int F_1 \, D_c \, d\Omega_c.$$

Let us calculate once and for all

$$n(x_1 y_1 z_1) = n_1 = N \int D_c \, d^3r_2 \, d^3r_3 \, \ldots \, d^3r_N$$

$$(d^3r_J = dx_J \, dy_J \, dz_J), \qquad (1)$$

where the last integration is not carried out. Then

$$\langle N_A \rangle = \int_V F_1 n_1 \, d^3r_1 = \int_A n_1 \, d^3r_1. \qquad (2)$$

n_1 is by definition the molecular density at the point x_1. From a "dimensional" point of view it is a number per volume. The condition (1) implies that the density satisfies the relation

$$N = \int_V n_1 \, d^3r_1. \qquad (3)$$

When the macroscopic medium is rigorously uniform the density is constant. Its variation from point to point in a crystalline solid is periodic in nature. In the thin interface between a liquid and the vapour above it varies very rapidly from point to point.

1.3. Double Density

Let us now calculate the mean of an expression

$$\sum_{J \neq K} \sum F_{JK},$$

$$\left[\text{or} \sum_{J \neq K} \sum F_{KJ} \quad \text{or} \quad \tfrac{1}{2} \sum_{J \neq K} \sum (F_{JK} + F_{KJ}) \right],$$

where F_{JK} is a function (not necessarily symmetrical) of the vectors r_J and r_K. This sum covers all the possible pairs of molecules and contains $N(N-1)$ terms. In the case when F_{JK} is symmetrical the number of different terms is only $\tfrac{1}{2}N(N-1)$. Nevertheless it is clear that

$$\left\langle \sum_{J \neq K} \sum F_{JK} \right\rangle = N(N-1) \langle F_{12} \rangle.$$

Let us calculate once and for all†

$$n(x_1 y_1 z_1, x_2 y_2 z_2)$$

† A volume integral without specified limits must be extended over the whole of the available volume.

5

or

$$n_{12} = N(N-1) \int D_c \, d^3r_3 \, d^3r_4 \ldots d^3r_N. \tag{1}$$

$n_{12} \, d^3r_1 \, d^3r_2$ is the probability of simultaneously finding the centre of some molecule in the element d^3r_1 and the centre of some other molecule in d^3r_2. The required average is

$$N(N-1) \langle F_{12} \rangle = \iint F_{12} n_{12} \, d^3r_1 \, d^3r_2, \tag{2}$$

where n_{12} will be called the double density. It is a symmetrical function. In accordance with formulae (1) and (3), n_1 can be calculated when we know n_{12}:

$$(N-1) \, n_1 = \int n_{12} \, d^3r_2. \tag{3}$$

A particularly simple statistical situation is one where the density in phase is a product of factors each of which depends only on the parameters of a single molecule:

$$D = f_1 f_2 \ldots f_N, \tag{4}$$

with

$$f_J = f(x_J, p_J).$$

The density in phase defined in this way satisfies the symmetry conditions correctly.

The normalization of D implies that of f:

$$\int f_J \, d^3r_J \, d^3p_J = 1. \tag{5}$$

In this case we shall say, provisionally, that the fluid is *statistically perfect*.

When the molecules can be treated as hard spheres and are therefore impenetrable the fluid cannot be perfect since formula (4) does not say that two molecules cannot occupy the same position. In a general way a fluid is never exactly perfect, but it is a question of a reference situation which on the one hand permits certain complete calculations and on the other hand constitutes a good approximation.

In the case of a perfect fluid we have:

$$n_1 = N \int f_1 \, d^3p_1$$

$$n_{12} = N(N-1) \int f_1 f_2 \, d^3p_1 \, d^3p_2,$$

6

and consequently

$$n_{12} = \left(1 - \frac{1}{N}\right) n_1 n_2. \tag{6}$$

When the molecules are numerous it is tempting to neglect $1/N$ here. We thus obtain the relation

$$n_{12} = n_1 n_2, \tag{7}$$

but it should be noted that this result contradicts formula (3). This kind of approximation, which consists basically of replacing $N(N - 1)$ by N^2, would eventually introduce more complications than simplifications. It is to be avoided.

Anyhow, let us consider two widely separated regions in the body of a fluid spread throughout a vast container: there are hardly any statistical relations between these two regions. The result is that when point 1 is taken from one of them and point 2 is taken from the other the approximation

$$n_{12} \approx n_1 n_2$$

is legitimate. But it should always be remembered that this relation cannot be absolutely rigorous and that it is best to apply it with discretion.

It should be stressed that the (approximate) absence of statistical connexions between distant regions is in no way imposed by any kind of statistical principle. An arbitrarily chosen density in phase D meets such a situation only exceptionally. It is a question of a physical phenomenon which will deserve serious examination.

Let us briefly generalize, to end this section, our definitions when particles of two kinds exist. There are N_a particles of kind a and N_b particles of kind b. We shall first be considering the simple densities

$$n_{a1} = N_a \int \ldots$$
$$n_{b1} = N_b \int \ldots, \tag{8}$$

then the double densities

$$n_{a12} = N_a(N_a - 1) \int \ldots$$
$$n_{ab12} = N_a N_b \qquad \int \ldots$$
$$n_{b12} = N_b(N_b - 1) \int \ldots. \tag{9}$$

7

The reader will be able to complete the expression of the integrals with ease. It is sufficient here to draw his attention to the fact that in general

$$n_{ab12} \quad \text{differs from} \quad n_{ab21},$$

whilst we always have

$$n_{ba21} = n_{ab12}.$$

1.4. Examples of Calculations of Averages

Let us assume that the molecules are centres of force which move in accordance with the laws of rational mechanics. The forces exerted on the molecules are of two kinds: on the one hand, forces exerted by an applied field in a medium made up by the molecules, for example the field of gravity; on the other hand, forces exerted by the molecules on each other. We shall assume that all these forces are independent of the velocity. The potential energy of the molecule in the applied field is

$$V(\mathbf{r}_J) \quad \text{or, simply,} \quad V_J.$$

This quantity may be time-dependent.

The mutual energies of the molecules are calculated in pairs. The pair J and K has a mutual energy

$$W(\mathbf{r}_J, \mathbf{r}_K) \quad \text{or} \quad W_{JK}$$

which will be assumed to depend only on the distance $|\mathbf{r}_K - \mathbf{r}_J|$. This being so, the mean potential energy of the system of molecules is clearly

$$\langle E_{\text{pot}} \rangle = \int n_1 V_1 \, d^3r_1 + \tfrac{1}{2} \int n_{12} W_{12} \, d^3r_1 \, d^3r_2. \tag{1}$$

The factor of $\tfrac{1}{2}$ corresponds in fact only to the pair JK; there is only one corresponding term in the expression for the potential energy, whilst the averaged double sum in the preceding section included two.

The hypothesis that the intermolecular forces are not dependent on the velocities and the hypothesis that these forces can be calculated in pairs are very restrictive. But few of the problems that have been touched on escape these limitations and we believe that at present the force model we are using is sufficient for discussing general properties. We should add that often the forces that we shall have to discuss will be "short-range" ones.

Another very simple example relates to the fluctuation of the number of molecules contained in a fraction A of the total volume V available for the movement of the molecules. We shall return to the notations which we used for defining the simple density in section 1.2. Let us recall that to say that a molecule is in a certain volume is to state, to use a more accurate expression, that its centre is in this volume.

The fluctuation in question is therefore

$$\Delta = N_A - \langle N_A \rangle, \qquad (2)$$

or

$$\Delta = \sum F_J - \int_A n_1 \, d^3r_1 \,.$$

Let us study its variance:

$$\langle \Delta^2 \rangle = \langle N_A^2 \rangle - (\langle N_A \rangle)^2 \,. \qquad (3)$$

Let us return to the expression

$$N_A = \sum F_J \,.$$

On squaring we note that, in accordance with the definition of F_J,

$$F_J^2 = F_J \,.$$

Therefore the double sum comprising $N(N-1)$ terms is

$$N_A^2 = \sum_J F_J + \sum \sum_{J \neq K} F_J F_K \,.$$

The mean value of this quantity is, in accordance with the preceding sections,

$$\langle N_A^2 \rangle = \int_A n_1 \, d^3r_1 + \int_A \int_A n_{12} \, d^3r_1 \, d^3r_2 \,.$$

Therefore,

$$\langle \Delta^2 \rangle = \int_A n_1 \, d^3r_1 + \int_A \int_A n_{12} \, d^3r_1 \, d^3r_2 - \left(\int_A n_1 \, d^3r_1 \right)^2 \,.$$

The last term can be written in the form of a double integral:

$$\left(\int_A n_1 \, d^3r_1 \right)^2 = \int_A \int_A n_1 n_2 \, d^3r_1 \, d^3r_2 \,.$$

This remark allows us to put the mean square of the fluctuation in a more elegant form:

$$\langle \Delta^2 \rangle = \int_A n_1 \, d^3r_1 + \int_A \int_A (n_{12} - n_1 n_2) \, d^3r_1 \, d^3r_2 \,. \qquad (4)$$

We shall return to this result on several occasions.

It could be shown by calculation that the variance of $N_{(V-A)}$ or of N_B is equal to that of N_A, but this result is obvious without calculations.

Let us apply formula (4) to the case when the fluid is perfect. In accordance with section 1.3 (6) we have

$$\langle \Delta^2 \rangle = \frac{\langle N_A \rangle \langle N_B \rangle}{N}.$$

If the volume A is small when compared with the total volume, N_B is very close to N and the result can be written more briefly and simply:

$$\langle \Delta^2 \rangle = \langle N_A \rangle.$$

This result is generally linked in the traditional treatment to the law of large numbers. The method which we have followed shows the matter in a simpler light: it is due solely to the hypothesis of "independence" and is exact even in a system which contains only two particles. There is good reason for not having recourse prematurely to the law of large numbers.

1.5. Microscopic Quantities

The molecular density n_1 was introduced as an indispensable intermediary in calculating the averages. But it is possible to find a quantity of which it is itself an average. Let us use in effect the Dirac δ distribution in three-dimensional space and consider a point r from the middle. The expression

$$v(r) = \sum_J \delta(r - r_J) \tag{1}$$

has the molecular density at the point r as its average. We shall say that this sum † represents the microscopic density at the point r. It is equally possible to find a quantity corresponding to the double density in the same way. We shall now consider two distinct points r and r'.

The double microscopic density is by definition

$$v(r, r') = \sum_{J \neq K} \sum \delta(r - r_J)\,\delta(r' - r_K). \tag{2}$$

† Massignon (1957) has extended this concept to velocity space.

10

It is easy to check that its mean value is the double molecular density.

Many other microscopic quantities can be thought of, for example, the intermolecular force exerted at the point r. By definition this will be

$$\zeta = \sum_{J \neq K} \sum \delta(r - r_J)\, F_{JK}. \tag{3}$$

F_{JK} denotes the force exerted by the molecule K on the molecule J. The force is derived from the mutual energy of the pair JK by the relation

$$F_{JK} = - \operatorname{grad}_J W_{JK}$$

which, by putting

$$r = |r_K - r_J|,$$

can also be written, in the central forces hypothesis, as

$$F_{JK} = \frac{r_K - r_J}{r} \frac{\mathrm{d}W_{JK}}{\mathrm{d}r}.$$

The mean value of the microscopic intermolecular force (at the point r_1) is

$$\langle \zeta_1 \rangle = \int n_{12} F_{12}\, \mathrm{d}^3 r_2. \tag{4}$$

The quotient of this quantity divided by the density at the same point is called the molecular field at this point:

$$F_1' = \frac{1}{n_1} \int n_{12} F_{12}\, \mathrm{d}^3 r_2. \tag{5}$$

When there is no correlation between the particles the molecular field is reduced to

$$F_1'' = \int n_2 F_{12}\, \mathrm{d}^3 r_2.$$

Whether there is any correlation or not this quantity always has a meaning. It is called the self-consistent field and is derived from the self-consistent potential:

$$V_1'' = \int n_2 W_{12}\, \mathrm{d}^3 r_2. \tag{6}$$

1.6. Intermolecular Pressure

Calculation of the pressure in a fluid proceeding from the molecular hypothesis presents many obscurities unless we have recourse to the idea of double density. It suffices to refer to Jeans' exciting

Dynamic Theory of Gases to see in the diversity of methods consecutively proposed that there was no guide line in the study of the question. And there was no question there of anything other than moderately dense gases. The problem seemed so much more complex to earlier theoreticians that they were led to treat separately the contribution from the repulsive forces, which are strong but have very short range, and from the attractive forces, which are weaker but have a longer range.

Following the concepts which are current nowadays, the problem is separated into two parts: first of all the pressure is expressed as a function of the densities and the intermolecular forces. This is a question where there is no need to state whether or not the fluid is in thermal equilibrium. It is only later that it is appropriate to introduce (if such is the situation) the properties of thermodynamic equilibrium while writing the density explicitly as a function of the state variables and of the intermolecular force law. In this way the calculations are sorted out and made more methodical.

This is the first part of the problem which we must treat. Let us consider some particular distribution of the molecules. Certain ones of them are in the fraction A of the total volume V. Let us again denote by F_J a function equal to 1 if J is in A and equal to 0 if it is not.

We assume that the forces are central; on the one hand

$$F_{JK} = -F_{KJ}, \tag{1}$$

whilst on the other hand F_{JK} is parallel to $r_K - r_J$. The resultant force to which molecules contained in A are subjected by all the other molecules, whether or not they are in A, is

$$\zeta_A = \sum_{J \neq K} \sum F_J F_{JK}. \tag{2}$$

The mean value of this sum in terms of the extension in phase can be expressed simply by means of the molecular field:

$$\langle \zeta_A \rangle = \int_A n_1 F_1' \, d^3 r_1. \tag{3}$$

On the other hand, since changing the name of the suffices does not change the value of a sum, ζ_A can also be written as

$$\zeta_A = \sum_{J \neq K} \sum F_K F_{KJ},$$

12

or, in accordance with formula (1),

$$\zeta_A = -\sum_{J \neq K}\sum F_K F_{JK}. \tag{4}$$

Half the sum of formulae (2) and of (4) gives:

$$\zeta_A = \tfrac{1}{2}\sum_{J \neq K}\sum (F_J - F_K) F_{JK}. \tag{5}$$

The factor $F_J - F_K$ is non-zero only when the pair JK is "riding" on the surface S which is the boundary of A. When the dimensions of this are great and there are pecularities on the molecular scale it is clear that the sum (3), by virtue of the short range of the intermolecular forces, has the appearance of a sum calculated along the surface. This is what we are about to define more precisely by means of a rigorous formalism: it will also be valid whatever the range of the forces or the structure of the surface. There is nevertheless a restriction: at each of its points the closed surface S which bounds the volume A has one and only one tangent plane.

We shall first estimate the quantity $F_K - F_J$. We consider some point P on the segment JK. The function F_P (in fact a distribution) has a derivative along this segment. We write

$$F_K - F_J = \int_J^K \frac{dF_P}{dr_P}\, dr_P, \tag{6}$$

noting that the integral jumps by $+1$ or -1 each time the point P crosses the surface S.

Next we use M to denote a point which for the moment belongs to the surface S and $\boldsymbol{\alpha}$ to denote the unit vector normal to the surface at this point. It points outwards. Finally we use $\boldsymbol{j}$ to denote the unit vector along JK. Formula (6) can be replaced by the following:

$$F_K - F_J = -\int_S \int_J^K (\boldsymbol{\alpha}.\boldsymbol{j})\, \delta(M - P)\, d^2 S_M\, dr_P. \tag{7}$$

This result is sophisticated but useful. It can be expressed as follows.

The point M now being arbitrary, the vector

$$-\boldsymbol{j}\int_J^K \delta(M - P)\, dr_P \tag{8}$$

is a function of this point, and its flux across the surface S is equal to the difference $F_K - F_J$.

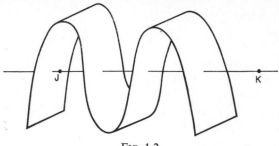

FIG. 1.2.

We thus obtain:

$$\langle \zeta_A \rangle = \int\limits_S \left\{ \left(\boldsymbol{\alpha} \cdot \tfrac{1}{2} \int n_{JK} \boldsymbol{j} \right) F_{JK} \times \left(\int\limits_J^K \delta(P - M) \, dr_P \right) d^3 r_J \, d^3 r_K \right\} d^2 S_M.$$

$$(9)$$

In this formula the point K describe all the space available, whilst the point P is subject to the straight line JK. It is preferable to look upon the point P as being free and to subject K to the straight line JP or to the straight line JM. We note that

$$d^3 r_K \, dr_P = \frac{r_{JK}^2}{r_{MJ}^2} \, d^3 r_P \, dr_{MK}.$$

The integral within the braces can be now written as

$$\frac{1}{2} \int \frac{1}{r_{JM}^2} \left(\int\limits_0^\infty n_{JK} r_{JK}^2 \boldsymbol{j} \, dr_{MK} \right) F_{JK} \, d^3 r_J \quad (\text{angle } \widehat{JMK} = 2\pi). \quad (10)$$

This expression can be made more symmetrical by using spherical coordinates centred on M:

$$d^3 r_J = r_{MJ}^2 \, dr_{MJ} \, d^2 O, \quad (11)$$

and if we remember that J, M and K are in a straight line, M being between J and K, expression (10) becomes

$$\tfrac{1}{2} \int\limits_{4\pi} \int\limits_0^\infty \int\limits_0^\infty n_{JK} r_{JK}^2 F_{JK} \boldsymbol{j} \, dr_{MJ} \, dr_{MK} \, d^2 O. \quad (12)$$

On the other hand, let us examine the tensorial aspect of the quantities obtained. We recall that

$$F_{JK} = \frac{r_K - r_J}{r_{JK}} \frac{dW_{JK}}{dr_{JK}}, \qquad j_x = \frac{x_K - x_J}{r_{JK}}.$$

14

The calculations have therefore brought to light a symmetrical second-rank tensor whose general component can be written as:

$$p_{Mxy} = -\frac{1}{2} \int_{4\pi} \int_0^\infty \int_0^\infty n_{JK}(x_K - x_J)(y_K - y_J)\frac{\mathrm{d}W}{\mathrm{d}r_{JK}}\,\mathrm{d}r_{MJ}\,\mathrm{d}r_{MK}\,\mathrm{d}^2O$$

$$(\widehat{JMK} = 2\pi). \tag{13}$$

This tensor is the tensor of the intermolecular pressure at the point M: the purpose of the sign convention adopted is definitely to have a pressure and not a tension.

This tensor can be used to express the mean flux of the forces exerted across the surface S by the external molecules on the internal molecules as follows:

$$\langle \zeta_A \rangle_x = -\int \alpha_y p_{xy}(M)\,\mathrm{d}^2 S_M. \tag{14}$$

The relation between the molecular field and the intermolecular pressure can finally be written:

$$nX' = -\frac{\partial p_{xx}}{\partial x} - \frac{\partial p_{xy}}{\partial y} - \frac{\partial p_{xz}}{\partial z}. \tag{15}$$

Formula (13) is rigorous but cumbersome. It is best to transform it for particular cases. We note first that if the forces are short-range ones and if the point M is far enough from the wall, which limits the system, the radial integrations are in fact limited by the range of the intermolecular forces.

Let us now assume in addition that the fluid is uniform in the vicinity of the point M; n_{JK} no longer depends independently on the points J and K but only on the vector $\mathbf{JK}$. Then

$$p_{xy} = -\frac{1}{2} \int_{4\pi} \int_0^\infty n_{JK}(x_K - x_J)(y_K - y_J) r_{JK}\frac{\mathrm{d}W}{\mathrm{d}r_{JK}}\,\mathrm{d}r_{JK}\,\mathrm{d}O. \tag{16}$$

If, finally, the fluid is isotropic the off-diagonal components are zero. The diagonal components are equal to one another. The intermolecular pressure is a scalar:

$$p_{sc} = -\frac{2\pi}{3} \int_0^\infty n_{JK}\frac{\mathrm{d}W}{\mathrm{d}r_{JK}} r_{JK}^3\,\mathrm{d}r_{JK}, \tag{17}$$

or, omitting the suffices that have become useless,

$$p = -\frac{2\pi}{3} \int\limits_0^\infty n_{12}(r)\frac{\mathrm{d}W}{\mathrm{d}r}r^3\,\mathrm{d}r. \tag{18}$$

This is the formula which allows us to estimate the intermolecular pressure in a uniform and isotropic fluid as a function of the double density and the potential of the intermolecular force. For these calculations to be valid all that is necessary is that the uniformity (and possibly the isotropy) should extend round the point M beyond the range of the intermolecular forces. It is satisfactory to find that in the case of short ranges the pressure at a point depends only on the structure of the fluid in the immediate vicinity. It can also be seen from formula (18) that the pressure is always positive when the force is repulsive.

It is useful to complete the results which have just been obtained for the case when the fluid is only approximately uniform. We shall first of all rewrite formula (13) in a less symmetrical way by replacing

$$\mathrm{d}r_{MJ}\,\mathrm{d}r_{MK} \quad \text{by} \quad \mathrm{d}r_{MJ}\,\mathrm{d}r_{JK}.$$

We obtain

$$p_{Mxy} = -\tfrac{1}{2}\int\limits_{4\pi}\int\limits_0^\infty\int\limits_0^{r_{JK}} n_{JK}j_x j_y r_{JK}^2(\mathrm{d}W/\mathrm{d}r_{JK})\,\mathrm{d}r_{MJ}\,\mathrm{d}r_{JK}\,\mathrm{d}^2O. \tag{19}$$

At the same time we shall write n_{JK}, which is a function of r_J and of r_K, as a function of

$$\frac{r_J + r_K}{2},\ r_{JK},\ j_x,$$

which can be written as

$$n_{JK} = f\!\left(\frac{r_J + r_K}{2},\ r_{JK},\ j_x\right). \tag{20}$$

As the double density is a symmetrical function of J and of K, f is an even function of the vector j.

The hypothesis that the medium is only approximately uniform leads to the following limit:

$$n_{JK} = f(r_M, r_{JK}, j_x) + \left(\frac{x_J + x_K}{2} - x_M\right)\frac{\partial f}{\partial x_M}$$

$$+ \frac{1}{2}\left(\frac{x_J + x_K}{2} - x_M\right)\left(\frac{y_J + y_K}{2} - y_M\right)\frac{\partial^2 f}{\partial x_M\,\partial y_M}, \tag{21}$$

16

or
$$n_{JK} = f(r_M, r_{JK}, j_x) + \frac{1}{2}(r_{MK} - r_{MJ})\, j_x \frac{\partial f}{\partial x_M}$$
$$+ \frac{1}{8}(r_{MJ} - r_{MK})^2\, j_x j_y \frac{\partial f^2}{\partial x_M\, \partial y_M}. \tag{22}$$

The first-order term makes no contribution to the pressure. On the other hand, integration over dr_{MJ} can be carried out. One obtains an expression where the principal term is that which has already been obtained in a perfectly uniform medium and a corrective term which brings in the second derivatives of f with respect to the M co-ordinates—we have not written out the suffices M, J and K explicitly here, and there is a tensorial summation over the x' and y':

$$p_{xy} = -\frac{1}{2} \int\limits_{4\pi} \int\limits_{0}^{\infty} f j_x j_y r^3 (dW/dr)\, dr\, d^2O$$
$$-\frac{1}{48} \frac{\partial^2}{\partial x'\, \partial y'} \int\limits_{4\pi} \int\limits_{0}^{\infty} f j_{x'} j_{y'} j_x j_y r^5 (dW/dr)\, dr\, d^2O. \tag{23}$$

In this result the summations over the three axes as far as the x' and y' suffices are concerned are understood as before.

The formalism can be readjusted a little. The following is another result relating to the average of the pressure in the available volume. Let us return to relation (15), which we can rewrite as follows:
$$\int X_{12} n_{12}\, d^3 r_2 = -\partial p_{1xy}/\partial y_1; \tag{24}$$
multiply both sides by $y_1\, d^3 r_1$, and then integrate. We obtain
$$\int y_1 X_{12} n_{12}\, d^3 r_1\, d^3 r_2 = -\int p_{1xy}\, d^3 r_1. \tag{25}$$
Then by virtue of symmetry
$$\int y_1 X_{12} n_{12}\, d^3 r_1\, d^3 r_2 = \int y_2 X_{21} n_{12}\, d^3 r_1\, d^3 r_2. \tag{26}$$
The final result is
$$\int p_{1xy}\, d^3 r_1 = -\tfrac{1}{2} \int (y_2 - y_1)\, X_{12} n_{12}\, d^3 r_1\, d^3 r_2. \tag{27}$$

This formula gives us an average idea of the pressure in the volume occupied by the medium without having recourse to formula (13), which may be useful when we do not wish to go into details, as in the case of a solid, for example. It is obvious that if the medium is practically uniform, formula (27) is the same as (16).

17

1.7. Liouville Equation

We have just been discussing the forces exerted in a system of molecules, but we have only considered the state of a system at a certain point in time. To give life to our formulae it is necessary now to follow the evolution of the system in the course of time under the influence of these forces.

We shall start at first from a very general point of view and we shall not restrict ourselves to a system of material points subject to central forces, but we shall discuss a more broadly defined system.

We shall stay within the framework of rational mechanics and we shall assume that the equations of the system's trajectory can take the Hamiltonian form. To make the formulae simpler we shall assume that the system has only two degrees of freedom (position x, momentum p). The Hamiltonian function $H(x, p, t)$ can then be used to calculate the motion:

$$\frac{\mathrm{d}x}{\mathrm{d}t} = \frac{\partial H}{\partial p}, \quad \frac{\mathrm{d}p}{\mathrm{d}t} = -\frac{\partial H}{\partial x}. \tag{1}$$

Let us consider a certain volume Ω of phase space at the time t. Each point P of this volume defines a trajectory. At the time t' the trajectory defined by P at the time t has reached the point P': when the point P describes the volume Ω, the point P' describes a volume Ω'. A first important point is that these volumes are equal. To show this it is best to argue with infinitesimal volumes and infinitesimal time intervals. The initial volume Ω will therefore be $\mathrm{d}x\,\mathrm{d}p$, the final volume $\mathrm{d}x'\,\mathrm{d}p'$ and the time interval δt. First of all, we have

$$x' = x + \frac{\partial H}{\partial p}\,\delta t,$$

$$p' = p - \frac{\partial H}{\partial x}\,\delta t.$$

As a result we obtain an expression for $\mathrm{d}x'$ and $\mathrm{d}p'$ as functions of $\mathrm{d}x$, $\mathrm{d}p$ and δt:

$$\mathrm{d}x' = \mathrm{d}x + \left(\frac{\partial^2 H}{\partial p\,\partial x}\,\mathrm{d}x + \frac{\partial^2 H}{\partial p^2}\,\mathrm{d}p\right)\delta t,$$

$$\mathrm{d}p' = \mathrm{d}p - \left(\frac{\partial^2 H}{\partial x^2}\,\mathrm{d}x + \frac{\partial^2 H}{\partial p\,\partial x}\,\mathrm{d}p\right)\delta t.$$

The relation between the elements of extension in phase is therefore

$$dx'\, dp' = \begin{vmatrix} 1 + \dfrac{\partial^2 H}{\partial p\, \partial x}\, \delta t & \dfrac{\partial^2 H}{\partial p^2}\, \delta t \\[2ex] -\dfrac{\partial^2 H}{\partial x^2}\, \delta t & 1 - \dfrac{\partial^2 H}{\partial p\, \partial x}\, \delta t \end{vmatrix} dx\, dp.$$

In expanding the determinant the first-order terms in δt disappear. In other words, we obtain the result that the volume Ω considered as a time function has an invariant measure

$$\frac{d\Omega}{dt} = 0.$$

This is the Liouville theorem (1838).† Let us now consider a probability distribution defined by the density in phase D. The probability of finding the system in the extension $d\Omega$ is

$$D\, d\Omega.$$

At time t' the system has evolved. If it started from the extension $d\Omega$ it is in the extension $d\Omega'$. This certainty brings with it a condition for the density in phase: the density in phase is such that the probability has not changed. We obtain

$$D\, d\Omega = D'\, d\Omega'.$$

But the above reasoning has shown us the constancy of the element of extension in phase along trajectories. The constancy of the probability means that the density in phase is equally constant along a trajectory:

$$\frac{dD}{dt} = 0.$$

We can rewrite this result by expressing the total time derivative in terms of partial derivatives:

$$\frac{\partial D}{\partial t} + \frac{\partial H}{\partial p}\frac{\partial D}{\partial x} - \frac{\partial H}{\partial x}\frac{\partial D}{\partial p} = 0. \qquad (2)$$

The result can be generalized for a more general system. Let x_1, p_1; x_2, p_2; ... be the set of pairs of canonically conjugate variables.

---

† The consequences of Liouville's theorem in statistical mechanics were introduced by Boltzmann (1871). It was J. W. Gibbs who showed the importance of Liouville's equation.

19

Probability Densities and their Evolution

The density in phase of the system evolves satisfying the equation

$$\frac{\partial D}{\partial t} + \sum \left(\frac{\partial H}{\partial p_i} \frac{\partial D}{\partial x_i} - \frac{\partial H}{\partial x_i} \frac{\partial D}{\partial p_i} \right) = 0. \tag{3}$$

This is the Liouville equation: it is fundamental. The formalism involves the Poisson brackets. Let us consider two physical quantities which are functions of the phase and time:

$$F(x_i p_i t),$$
$$G(x_i p_i t).$$

The Poisson bracket of F and G is by definition the expression

$$[F, G] = \sum_i \left[\frac{\partial F}{\partial x_i} \frac{\partial G}{\partial p_i} - \frac{\partial F}{\partial p_i} \frac{\partial G}{\partial x_i} \right]. \tag{4}$$

Let us now calculate the derivative of the quantity F along a trajectory described by the system:

$$\frac{dF}{dt} = \frac{\partial F}{\partial t} + \sum_i \left[\frac{dx_i}{dt} \frac{\partial F}{\partial x_i} + \frac{dp_i}{dt} \frac{\partial F}{\partial p_i} \right]. \tag{5}$$

By virtue of the equations of motion (1) this can also be written as

$$\frac{dF}{dt} = \frac{\partial F}{\partial t} - [H, F]. \tag{6}$$

We shall adopt the following notation for the particular Poisson bracket formed by the Hamiltonian and phase function:

$$\mathscr{L}F = - [H, F]. \tag{7}$$

The operator $\mathscr{L}$ introduced here will be called the Liouville operator and the Liouville equation will be written in the condensed form

$$\frac{\partial D}{\partial t} + \mathscr{L}D = 0. \tag{8}$$

The Liouville equation † is a first-order, linear, and homogeneous equation. A certain number of obvious consequences result from it. In particular, if D and D' are two-phase functions which satisfy the Liouville equation, then their product DD' also satisfies it. It should be clear that here we are considering functions which have

† In applications it is often possible and convenient to take the velocities as independent variables instead of the momenta (cf. Delcroix, 1963).

derivatives insofar as that is necessary. If D satisfies the Liouville equation, then the same can be said of $f(D)$ and of D^2. The Liouville equation, therefore, has solutions that are never negative. If D is a solution which at a certain point in time is not negative it keeps this property indefinitely. This is a result of the fact that D satisfies the equation

$$\frac{\mathrm{d}D}{\mathrm{d}t} = 0.$$

These remarks are essential since D is capable of representing a probability density.

When the Hamiltonian function H is independent of time it is obviously a solution of the Liouville equation since

$$[H, H] = 0.$$

This does not mean that H could represent a density in phase. We should add that more generally any function of H likewise satisfies the Liouville equation. The Liouville equation can be integrated formally as follows:

$$D(t) = \mathrm{e}^{-\mathscr{L}(t - t_0)} D(t_0).$$

We shall not be making much use of this formula which obliges us to consider the exponential of an operator, but its quantum generalization will be of great help to us. We assumed here that $\mathscr{L}$ was not explicitly time-dependent.†

Let us now find the Liouville equation for a system of N molecules subject to a force field and exerting on each other central forces that are a function of the distance only. Knowing that

$$H = \sum_J \frac{1}{2m} p_J^2 + \sum_J V_J + \frac{1}{2} \sum_{J \neq K} W_{JK},$$

it is easy to obtain the following result:

$$\frac{\partial D}{\partial t} + \sum_J \left(\frac{p_J}{m} \frac{\partial}{\partial x_J} + (X_J + \sum_K X_{JK}) \frac{\partial}{\partial p_J} \right) D = 0. \qquad (9)$$

† The question of the integration of the Liouville equation over a long period of time is part of the ergodic theory. The problem of the relations between ergodic theory and statistical mechanics has been reviewed recently by I. E. Farquhar (1964). I must admit that I share his disillusioned opinion (p. 199).

The notation has been abbreviated.† For example, $p(\partial/\partial x)$ stands for

$$p\frac{\partial}{\partial x} + q\frac{\partial}{\partial y} + r\frac{\partial}{\partial z}.$$

The most immediate generalization of equation (9) relates to the case when there is a magnetic field and the particles carry an electric charge.

1.8. Time Derivatives

As an application of the formalism we shall calculate the time derivative of the average of a quantity F. This average, by definition, is

$$\langle F \rangle = \int DF \, d\Omega. \tag{1}$$

The time derivative can be calculated directly:

$$\frac{d}{dt}\langle F \rangle = \int \left(\frac{\partial D}{\partial t} F + D\frac{\partial F}{\partial t}\right) d\Omega.$$

We eliminate the derivative of the density in phase using the Liouville equation:

$$\frac{d}{dt}\langle F \rangle = \int \left(-F\mathscr{L}D + D\frac{\partial F}{\partial t}\right) d\Omega. \tag{2}$$

The operator $\mathscr{L}$ has the following property:

$$\mathscr{L}FD = F\mathscr{L}D + D\mathscr{L}F.$$

We obtain

$$\frac{d}{dt}\langle F \rangle = \int \left[D\left(\frac{\partial F}{\partial t} + \mathscr{L}F\right) - \mathscr{L}DF\right] d\Omega. \tag{3}$$

But, according to the rules of operator calculus

$$\frac{\partial H}{\partial p_i}\frac{\partial}{\partial x_i} - \frac{\partial H}{\partial x_i}\frac{\partial}{\partial p_i} = \frac{\partial}{\partial x_i}\left(\frac{\partial H}{\partial p_i}\right) - \frac{\partial}{\partial p_i}\left(\frac{\partial H}{\partial x_i}\right).$$

The volume integral

$$\int \mathscr{L}DF \, d\Omega,$$

which extends over the whole accessible region of phase space, can be evaluated just as well by a flux across the boundary (which

† See also Note on Notation on p. xix.

can be pushed to infinity) of this region. The surface integration will no longer introduce any derivative of the density in phase, but the density in phase itself via the products

$$D\frac{\partial H}{\partial p_i} \quad \text{and} \quad D\frac{\partial H}{\partial x_i}.$$

There is an important set of systems for which the surface integral is zero. These systems will be called "closed". It is difficult to give a definition of them that is really satisfactory and general. The "closing" of a system is connected on the one hand with the mechanical laws to which it is subject and on the other hand to conditions imposed on the density in phase. These conditions must, of course, be invariant in the course of time. To make this clear let us consider a system made up of molecules that are force centres. If the molecules are enclosed in a container we can picture mathematically the physical role of the container by imagining that an infinitely strong repulsive field is exerted at its surface which prevents the molecules from leaving. No molecule will be able to reach the boundary wall without acquiring an infinite potential energy. But such a situation will be impossible if we admit that the density in phase tends rapidly towards zero when the system's energy becomes very great. Therefore the density in phase will be zero at the boundary of the configuration space. The same will be true for momentum space. Unless the system is enclosed in a container we can admit that the probability that some molecule is a long way from the origin is zero. Since the molecules have only finite velocities this property is qualitatively preserved indefinitely and the closing will be achieved.

The problem of defining a closed system is similar to the one in quantum mechanics of defining a Hermitian operator: the Hermitian property is connected on the one hand with the structure of the operator, but equally with the properties imposed on the wave function at the boundary of the region where it exists. It will also be noted that when a system is closed the operator $i\mathscr{L}$ ($i = \sqrt{-1}$) that corresponds to it is Hermitian.

However, when a system is closed the equation (3) is simplified and takes the following form:

$$\frac{\mathrm{d}}{\mathrm{d}t}\langle F\rangle = \left\langle \frac{\mathrm{d}F}{\mathrm{d}t}\right\rangle. \tag{4}$$

23

This result can be expressed as follows: the time derivative of the average of a quantity is equal to the average of the time derivative of this quantity. This is why the derivative of the average position of a molecule is equal to the average of its velocity:

$$\frac{d}{dt} \langle r_J \rangle = \langle c_J \rangle.$$

It is easy to think of other simple examples. The simplest case is that of the average of 1. This average is the integral

$$\int D \, d\Omega$$

and its time derivative is clearly zero in accordance with the preceding theorem. The density in phase normalized at a given point in time keeps this property indefinitely. This result was necessary for it to keep indefinitely one of the essential characteristics of a probability density. If F is a function of the Hamiltonian (it is assumed that the latter is independent of time) its derivative is likewise zero.

1.9. Chain of Recurrence Equations of Motion

The densities we discussed in section 1.2 related only to configuration space. We now need less restrictive densities relating to both the position and the momentum of one molecule (simple density) or of two molecules (double density). We shall denote by $d\Omega_J$ the phase space element relative to the molecule J. Thus we define the simple density in phase as

$$\mu(r_1, p_1) \quad \text{or} \quad \mu_1 = N \int D \, d\Omega_2 \, d\Omega_3 \ldots d\Omega_N,$$

and the analogous double density as

$$\mu(r_1, p_1, r_2, p_2) \quad \text{or} \quad \mu_{12} = N(N-1) \int D \, d\Omega_3 \, d\Omega_4 \ldots d\Omega_N.$$

Some relations are trivial:

$$\left.\begin{array}{l} n_1 = \int \mu_1 \, d^3 p_1, \\[4pt] n_{12} = \int \mu_{12} \, d^3 p_1 \, d^3 p_2, \\[4pt] (N-1)\, \mu_1 = \int \mu_{12} \, d\Omega_2. \end{array}\right\} \tag{1}$$

These quantities n_1, n_{12}, μ_1, μ_{12}, which depend on only a limited number of parameters, are given the name of reduced quantities. Let us multiply both sides of the Liouville equation by

$$N \, d\Omega_2 \, d\Omega_3 \, ..., d\Omega_N,$$

and integrate over the available phase space. All the terms where a derivative allows us to reduce the number of successive integrations by one lead to a zero result by virtue of the hypothesis that the system is closed. On the other hand, numerous terms are equal because of the symmetry of the density in phase with respect to the molecules. Finally, returning to equation (9) of section 1.7, we arrive at

$$\frac{\partial \mu_1}{\partial t} + \frac{p_1}{m} \frac{\partial \mu_1}{\partial x_1} + X_1 \frac{\partial \mu_1}{\partial p_1} + \int X_{12} \frac{\partial \mu_{12}}{\partial p_1} d\Omega_2 = 0. \qquad (2)$$

If there was no intermolecular force, all we should obtain here would be the Liouville equation for a system containing only one particle. If the forces are short-range ones the integral figuring in equation (2) covers only the immediate vicinity of r_1. The properties at each point in a medium where the molecules have properties of this kind are connected only with the state of the surrounding area. This is what happens in hydrodynamics as opposed to the case of electrodynamics. In accordance with equation (2) the evolution of the simple density depends on the double density. In the same way the evolution of the double density in phase depends on the triple density:

$$\frac{\partial \mu_{12}}{\partial t} + \frac{p_1}{m} \frac{\partial \mu_{12}}{\partial x_1} + \frac{p_2}{m} \frac{\partial \mu_{12}}{\partial x_2} + (X_1 + X_{12}) \frac{\partial \mu_{12}}{\partial p_1}$$

$$+ (X_2 + X_{21}) \frac{\partial \mu_{12}}{\partial p_2}$$

$$+ \int \left(X_{13} \frac{\partial \mu_{123}}{\partial p_1} + X_{23} \frac{\partial \mu_{123}}{\partial p_2} \right) d\Omega_3 = 0. \qquad (3)$$

The transition from (3) to (2) is simple. All that we have to do is to multiply by $d\Omega_2$, integrate and apply some fundamental formulae. This method, which reduces the number of independent variables and allows us to pass from a quantity that is exact but hard to manipulate because of the parameters it contains to a poorer but simpler quantity, may be called the regressive method.

25

Probability Densities and their Evolution

Higher-order equations could easily be written out. The series of equations (2), (3), etc., is the chain of recurrence equations of motion.

It should be noted that it will be useful for us to consider higher-order reduced densities and even to go as far as

$$\mu_{1...N} = N!D.$$

This last density, of course, hardly deserves the name "reduced".

1.10. Diffusion Equations

We shall study in turn the diffusion of the molecules, the diffusion of the momentum and the diffusion of the energy. We must first of all introduce the velocity distribution function.

Momentum distribution function

Let us put

$$\mu_1 = n_1 f_1. \tag{1}$$

In this breakdown of the simple density in phase as a product the first factor is the molecular density: it depends only on the spatial coordinates. On the other hand the second factor depends in general both on these coordinates and the momenta: it is the momentum distribution function. It satisfies the normalization condition

$$\int f_1 \, d^3 p_1 = 1. \tag{2}$$

Let us now consider some function F of only the momentum p; such a function will, in general, represent a physical quantity relating to a single particle. We shall calculate

$$\langle F_1 \rangle = \int F_1 f_1 \, d^3 p_1. \tag{3}$$

The average obtained still depends on the spatial coordinates. It is a "local" average. Some of these averages play an important part and will be used in the following sections. They are convenient intermediaries for calculating the true averages, the overall averages. In fact

$$\int n_1 \langle F_1 \rangle \, d^3 r_1 = \left\langle \sum_J F_J \right\rangle. \tag{4}$$

The simplest case is the one where F is the same as the momentum itself. Thus this latter has a local average

$$\langle p_1 \rangle = \int p_1 f_1 \, d^3 p_1, \tag{5}$$

which of course corresponds to a local velocity

$$\langle c_1 \rangle = \frac{1}{m} \langle p_1 \rangle. \tag{6}$$

This formula defines at any point in time a velocity field which is the hydrodynamic velocity field.

Having stated this, it is advisable, so as to be able to study other averages of the same kind, to consider the fluctuation of the momentum

$$P_1 = p_1 - \langle p_1 \rangle \tag{7}$$

whose average is zero. It is in this way that we shall express the local kinetic energy

$$\frac{1}{2m} \langle p_1^2 \rangle = \frac{1}{2m} [\langle p_1 \rangle^2 + \langle P_1^2 \rangle] \tag{8}$$

which is a sum of the hydrodynamic kinetic energy and thermal kinetic energy. Likewise

$$\langle p_1 q_1 \rangle = \langle p_1 \rangle \langle q_1 \rangle + \langle P_1 Q_1 \rangle. \tag{9}$$

The local fluctuation of the velocity will likewise be

$$C_1 = c_1 - \langle c_1 \rangle. \tag{10}$$

Its components will be denoted by U_1, V_1, W_1 (to avoid any confusion the potential energy of the applied field will in future be denoted by Φ_j).

We propose now to calculate the time derivatives of the principal local averages.

We shall proceed from the first recurrence equation which we rewrite as

$$\frac{\partial}{\partial t} \mu_1 + \left(\frac{p_1}{m} \frac{\partial}{\partial x_1} + X_1 \frac{\partial}{\partial p_1} \right) \mu_1 + \int X_{12} \frac{\partial}{\partial p_1} \mu_{12} \, d\Omega_2 = 0. \tag{11}$$

Particle diffusion

The derivative of the molecular density can be obtained by multiplying both sides of the preceding equation by $d^3 p_1$ and by integrating over momentum space. The terms that contain a derivative of the momentum give a zero result. By suppressing the suffix 1 we obtain

$$\frac{\partial}{\partial t} n + \frac{\partial}{\partial x} n \, \langle u \rangle = 0. \tag{12}$$

This equation expresses how the molecules diffuse because of hydro-dynamic motion. This is a conservation equation, which is also called the equation of continuity.

Momentum diffusion†

The equation is now multiplied by p_1. Integrations by parts give the terms containing $\partial/\partial p_1$ the appropriate form. We obtain

$$\frac{\partial}{\partial t} n_1 \langle p_1 \rangle + \frac{\partial}{\partial y_1} n_1 \frac{\langle p_1 \rangle \langle q_1 \rangle + \langle P_1 Q_1 \rangle}{m}$$

$$= n_1 X_1 + \int n_{12} X_{12} \, d^3 r_2. \tag{13}$$

We shall express the last integral as a function of the intermolecular pressure

$$\int n_{12} X_{12} \, d^3 r_2 = - \frac{\partial}{\partial y_1} p_{1xy}. \tag{14}$$

We shall define a kinetic pressure

$$K_{1xy} = n_1 \frac{\langle P_1 Q_1 \rangle}{m}. \tag{15}$$

This pressure is a symmetrical quantity with nine components whose diagonal terms are all positive. We finally define the total pressure

$$P_{1xy} = K_{1xy} + p_{1xy}, \tag{16}$$

which is the sum of the kinetic pressure and the intermolecular pressure. We now obtain

$$\frac{\partial}{\partial t} n_1 \langle p_1 \rangle + \frac{\partial}{\partial y_1} n_1 \frac{\langle p_1 \rangle \langle q_1 \rangle}{m} = n_1 X_1 - \frac{\partial}{\partial y_1} P_{1xy}. \tag{17}$$

This equation can be put in a more traditional form by substituting the hydrodynamic velocity for the local momentum and by eli-

---

† J. H. Irving and J. G. Kirkwood (1950). In the case of the energy the results depend a little on the rigour one tries to apply. In addition the conventions that can be adopted are to some extent arbitrary.

minating $$\partial n_1/\partial t$$

using the conservation equation. Dropping the suffix 1 we obtain

$$m\left[\frac{\partial}{dt} + (v)\frac{\partial}{\partial y}\right](u) = X - \frac{1}{n}\frac{\partial}{\partial y}P_{xy}. \tag{18}$$

This is the general form of the hydrodynamic equation. It is general in the sense that there is nothing in it that specifies the properties of the total pressure.

From the way this equation describes phenomena observed in hydrodynamics or hydrostatics we can deduce that the overall pressure introduced here is the pressure measured by manometers and similar instruments.

If the forces are long-range ones the self-consistent field must be separated from the intermolecular pressure to add it to the applied field.†

It is often convenient to denote the operator

$$\frac{\partial}{\partial t} + (v_1)\frac{\partial}{\partial y_1} \quad \text{by} \quad \frac{d}{dt}. \tag{19}$$

Energy diffusion

This subsection is limited to the case of short-range forces.

The Hamiltonian of the system of point particles can be written as

$$H = \sum_J\left(\frac{p_J^2}{2m} + \Phi_J\right) + \frac{1}{2}\sum_J\sum_K W_{JK}. \tag{20}$$

The pair JK makes a contribution W_{JK} to this energy. We wish to be able to define the energy in a partial volume A. The kinetic energy and the energy in the applied field do not cause us any difficulty here, but the intermolecular energy, which is not connected with a definite particle, is embarrassing.

We shall share, rather arbitrarily, the energy W_{JK} between the two molecules J and K, attributing half to each. We can then speak of the energy per particle

$$H_J = \frac{1}{2m}p_J^2 + \Phi_J + \frac{1}{2}\sum_K W_{JK}. \tag{21}$$

† See section 2.9.

We put

$$\langle H_1\rangle = \frac{1}{2m}\,\langle p_1^2\rangle + \Phi_1 + \frac{1}{2n_1}\int n_{12}W_{12}\,\mathrm{d}^3r_2. \qquad (22)$$

1. We replace $\langle p_1^2 q_1\rangle$ by

$$\langle p_1^2\rangle\,\langle q_1\rangle + \langle p_1^2 Q_1\rangle,$$

then $\langle p_1^2 Q_1\rangle$ by

$$\langle p_1^2\rangle\,\langle Q_1\rangle + 2\langle p_1\rangle\,\langle P_1 Q_1\rangle + \langle P_1^2 Q_1\rangle, \qquad (23)$$

an expression where the first term is zero.

We replace $\int \mu_{12}u_1 X_{12}\,\mathrm{d}^3p_1\,\mathrm{d}\Omega_2$ by

$$\int \mu_{12}[\langle u_1\rangle + U_1]\,X_{12}\,\mathrm{d}^3p_1\,\mathrm{d}\Omega_2, \qquad (24)$$

or by

$$\langle u_1\rangle\int n_{12}X_{12}\,\mathrm{d}\Omega_2 + \int \mu_{12}U_1 X_{12}\,\mathrm{d}^3p_1\,\mathrm{d}\Omega_2. \qquad (25)$$

We regroup the terms to make the pressure appear explicitly:

$$n_1\,\frac{\mathrm{d}}{\mathrm{d}t}\,\frac{\langle p_1^2\rangle}{2m} = n_1\,\langle u_1\rangle\,X_1 - \frac{\partial}{\partial y_1}\,\langle u_1\rangle\,K_{1xy} - \langle u_1\rangle\,\frac{\partial p_{1xy}}{\partial y_1}$$
$$- \frac{\partial}{\partial y_1}\,n_1\,\frac{\langle P_1^2 Q_1\rangle}{2m^2} + \int \mu_{12}U_1 X_{12}\,\mathrm{d}^3p_1\,\mathrm{d}\Omega_2. \qquad (26)$$

2. The derivative of the second term can be obtained directly by means of the conservation equation

$$\frac{\partial}{\partial t}\,n_1\Phi_1 = -\Phi_1\,\frac{\partial}{\partial y_1}\,\langle v_1\rangle\,n_1 + n_1\,\frac{\partial\Phi_1}{\partial t}, \qquad (27)$$

or, by introducing the applied force,

$$n_1\,\frac{\mathrm{d}}{\mathrm{d}t}\,\Phi_1 = -\,n_1\,\langle u_1\rangle\,X_1 + n_1\,\frac{\partial\Phi_1}{\partial t}. \qquad (28)$$

3. The derivative of n_{12} can be obtained by means of the second recurrence equation

$$\frac{\partial n_{12}}{\partial t} = -\int\!\left(\frac{\partial}{\partial x_1}\,u_1\mu_{12} + \frac{\partial}{\partial x_2}\,u_2\mu_{12}\right)\mathrm{d}^3p_1\,\mathrm{d}^3p_2, \qquad (29)$$

or, introducing the local average of **c**,

$$\frac{\partial n_{12}}{\partial t} = -\frac{\partial}{\partial x_1} \langle u_1 \rangle\, n_{12} - \frac{\partial}{\partial x_2} \langle u_2 \rangle\, n_{12}$$

$$- \int\left(\frac{\partial}{\partial x_1} U_1\mu_{12} + \frac{\partial}{\partial x_2} U_2\mu_{12}\right) d^3p_1\, d^3p_2. \quad (30)$$

The derivative of the third term can be deduced from this:

$$\frac{\partial}{\partial t}\int \frac{1}{2} n_{12} W_{12}\, d^3r_2$$

$$= -\int \frac{1}{2} W_{12}\left[\frac{\partial}{\partial x_1}\langle u_1 \rangle\, n_{12} + \frac{\partial}{\partial x_2}\langle u_2 \rangle\, n_{12}\right] d^3r_2$$

$$- \int \frac{1}{2} W_{12}\left(\frac{\partial}{\partial x_1} U_1\mu_{12} + \frac{\partial}{\partial x_2} U_2\mu_{12}\right) d^3p_1\, d\Omega_2. \quad (31)$$

In the second member we integrate by parts, obtaining

$$\left[\frac{\partial}{\partial t} + \frac{\partial}{\partial y_1}\langle v_1 \rangle\right]\int \frac{1}{2} n_{12} W_{12}\, d^3r_2$$

$$= -\frac{1}{2}\langle u_1 \rangle \int n_{12} X_{12}\, d^3r_2 + \frac{1}{2}\int \langle u_2 \rangle\, n_{12} X_{12}\, d^3r_2$$

$$- \frac{\partial}{\partial x_1}\int \frac{1}{2} U_1 W_{12}\mu_{12}\, d^3p_1\, d\Omega_2$$

$$- \frac{1}{2}\int (U_1 - U_2)\, X_{12}\mu_{12}\, d^3p_1\, d\Omega_2. \quad (32)$$

We shall simplify matters a little by assuming that in the region in question there is little variation from point to point in the hydrodynamic velocity in the region within the range of the forces; this will allow us in (32) to replace

$$\langle u_2 \rangle \quad \text{by} \quad \langle u_1 \rangle + (y_2 - y_1)\frac{\partial\langle u_1 \rangle}{\partial y_1}$$

and consequently the first two integrals on the right-hand side by

$$\frac{\partial\langle u_1 \rangle}{\partial y_1}\int \frac{1}{2} n_{12}(y_2 - y_1)\, X_{12}\, d^3r_2.$$

Let us assume, at the same time as our hypothesis about the velocities, that the medium is approximately uniform. Then the above integral is the kinetic pressure with opposite sign

$$p_{1xy} \approx -\tfrac{1}{2}\int n_{12}(y_2 - y_1)\, X_{12}\, d^3r_2. \quad (33)$$

4*

Bringing all the terms together we obtain

$$n_1 \frac{\mathrm{d}}{\mathrm{d}t} \langle H_1 \rangle = n_1 \frac{\partial \Phi_1}{\partial t} - \frac{\partial}{\partial y_1} \langle u_1 \rangle P_{1xy}$$

$$- \frac{\partial}{\partial y_1} n_1 \frac{\langle P_1^2 Q_1 \rangle}{2m^2} - \frac{\partial}{\partial y_1} \int \frac{1}{2} V_1 W_{12} \mu_{12} \, \mathrm{d}^3 p_1 \, \mathrm{d}\Omega_2$$

$$+ \frac{1}{2} \int (U_1 + U_2) X_{12} \mu_{12} \, \mathrm{d}^3 p_1 \, \mathrm{d}\Omega_2 . \qquad (34)$$

The last integral, apart from the presence of the factor $U_1 + U_2$, resembles the integral which was used to define the intermolecular pressure. As in the case of the intermolecular pressure this integral contains an odd integrand. This means that it changes sign when we permute 1 and 2. For the same reasons as in the case of the pressure this integral can be put in the form of a derivative, but of a vector this time instead of a tensor. It is sufficient for us to give the expression for the case of a quasi-uniform medium. Then we can write

$$\frac{1}{2} \int (U_1 + U_2) X_{12} \mu_{12} \, \mathrm{d}^3 p_1 \, \mathrm{d}\Omega_2$$

$$= \frac{\partial}{\partial y_1} \frac{1}{4} \int (U_1 + U_2)(y_2 - y_1) X_{12} \mu_{12} \, \mathrm{d}^3 p_1 \, \mathrm{d}\Omega_2 . \qquad (35)$$

Let us thus put

$$W_{1x} = n_1 \frac{\langle Q_1^2 P_1 \rangle}{2m^2} + \frac{1}{2} \int U_1 W_{12} \mu_{12} \, \mathrm{d}^3 p_1 \, \mathrm{d}\Omega_2$$

$$- \frac{1}{4} \int (V_1 + V_2)(x_2 - x_1) Y_{12} \mu_{12} \, \mathrm{d}^3 p_1 \, \mathrm{d}\Omega_2 . \qquad (36)$$

The energy diffusion equation can finally be put in the form

$$n \frac{\mathrm{d}}{\mathrm{d}t} \langle H \rangle = n \frac{\partial \Phi}{\partial t} - \frac{\partial}{\partial y} \langle u \rangle P_{xy} - \frac{\partial}{\partial x} W_x . \qquad (37)$$

We can see in this equation

(1) the effect of hydrodynamic motion,
(2) the effect of time variations of the applied potential,
(3) the effect of pressures,
(4) a last effect which is normally described as the effect of heat losses.

The vector W is therefore the heat flux vector. The heat is defined here as in thermodynamics: it is what is missing in the energy balance when account has been taken of the effect of pressure. Statistical mechanics shows no originality here. Heat defined in this way is conservative when the applied field is independent of time and when the hydrodynamic velocity is zero, as is appropriate.

Conclusions

Our diffusion equations, conservation equation, equation of motion, and energy diffusion equation make up the hydrodynamic equations.

This set does not constitute a closed system which can be integrated with respect to time since, as well as the quantities whose derivative has been given, i.e. density, mean velocity and mean energy, it contains auxiliary quantities—pressure and heat flux.

Hydrodynamic and thermodynamic experts make up these deficiencies on the basis of experimental data. In accordance with their needs they give expressions for the auxiliary quantities that are more or less refined, but which permit a closed system to be written in every case. The diversity of actual cases leads to a whole collection of formulae which are more or less accurate, but which are included in the scheme which we have been discussing.

The part of the theoretician, into which we shall not go deeply in this volume, is to explain why we can describe natural phenomena by means of equations which are much less complex than the set of recurrence equations and then to justify or explain the elimination of the quantities we have called auxiliary. A start is given to this by the study of thermal equilibrium in Chapter 4 and the following chapters.

The theory is inevitably rather schematic. Hydrodynamic, thermodynamic and acoustic experts do not limit their investigations to our systems of rudimentary particles. Their equations extend to the case of complex molecules (oxygen, nitrogen, water), to the case of mixtures like air of liquid metals, and more generally of ionized media.

The case of strongly ionized gases or plasmas embarrasses everyone: the theoretician because the Coulomb forces (long-range forces) present difficulties which are not there with short-range forces, and the experimenter because of the instabilities that occur.

33

It is probable that the comfortable set-up of hydrodynamics or aerodynamics is only of limited application here. In addition a new agent, the magnetic field, appears here. Hydrodynamics as it expands becomes magnetohydrodynamics.

The at least partial failure of the hydrodynamic framework in the case of plasmas leads us to ask the question of whether we know if this framework is even satisfactory for the more commonplace fluids. All we can mention here are the problems connected with large density gradients and shock waves, the problem of the conditions at the walls, and that of low densities.

We should perhaps also mention the question of turbulence here. We can describe a turbulent fluid only thanks to statistical considerations, but it is a question of relatively macroscopic statistics rather than molecular statistics. Those who are anxious to synthesize may have the impression that these are one set of statistics too many. The attempts at reduction that have been made in this direction have preserved an academic nature up to the present. But this is perhaps because exactly neutral fluids lend themselves but little to any attempt at unification. It is possible that plasmas, in which rest is more of an anomaly, will make us take a fresh look at the problem.

Our description is based on classical mechanics. Reality is quantal. The classical description keeps its value for two reasons, at least for the fluids that we have quoted: the first, which is negative, is that the quantum equivalent of classical hydrodynamics is at a rather anaemic stage of development in this year of 1968, and the second is that less difference is shown between the applications of our two types of mechanics in the macroscopic realm than in the atomic field.

On the other hand, liquid helium and electrons in metals display strong quantum effects.

We note finally that real particles emit, absorb, or scatter radiation and that to the study of mixtures of molecules with other molecules we must add the study of mixtures of particles and photons.

CHAPTER 2

Occupation

2.1. Uncertainty in the Number of Particles Present

Let us continue to study the case where the fluid under study contains particles of only one kind. All our arguments up to date have been based on an essential datum—the number of particles making up the fluid. In fact, if we are dealing with a macroscopic system we can no more determine the exact number of particles it contains, which we shall call its occupation, than its initial state. It is therefore quite natural to bring this new factor of uncertainty into play in our arguments. As long as it is a question of indestructible particles this factor is not fundamental but it has many practical advantages to offer.

We shall denote the probability that the system contains N particles by π_N. N can take up any of the values 0, 1, ..., N, ...

If the particles are simple point particles the number N has no upper limit. If we assume that they have a hard core this number is limited by the dimensions of the container. The limit can be determined by considering the densest possible packing; it can be determined exactly if the container is one of certain simple shapes and with less accuracy if it is not. In fact, we shall find it convenient, as we have already said, to ascribe a probability to impossible situations, the corresponding probability being of course zero.

The set of π_N is normalized by the following condition:

$$\sum_N \pi_N = 1.$$

The average number of particles contained in the system, or the average occupation number or content can be written as

$$\langle N \rangle = \sum_N \pi_N N.$$

It is only exceptionally that this number is a whole number.

In studying macroscopic problems the content is very large and is of the order of magnitude of the Avogadro number. Certain approximations are therefore justified. In any case we shall derive our fundamental formulae without using this possibility.

That there can be uncertainty in the occupation number results from its being impossible to determine this quantity experimentally. Weighing procedures, for example, however perfect they may be, always leave a margin of uncertainty. Other factors of a different physical nature also play an equivalent part: inaccuracies in the dimensions of the containers, accidental adsorptions, impurities. Even the repetition of a measurement with the same equipment carefully isolated is not protected from the perturbations which are equivalent to a change in the occupation.

When the occupation number is high it might seem supererogatory to pay any interest to probabilities which correspond to very small occupations and in particular to π_0, the probability of zero occupation. Nevertheless we should note, in order to justify its consideration, that the short history of science has recorded cases when the operator had forgotten to fill his equipment with the required product. The reader will find that we are considering extreme cases here: we would remind him, so as not to seem to be attacking only the honour of physicists here, of his gastronomic experience; he must have eaten soups or dishes which the specialist had forgotten to salt. From the theoretical point of view the reader will admit that the quantity π_0, even if it is immeasurably small, is a very useful quantity to know.

2.2. Densities

When the system contains N particles the corresponding density in phase will be denoted by $D_{12...N}$. This is the density in phase and of occupation. Several ways can be thought of to normalize this quantity. The most convenient one consists in insisting that the integration in the extension in phase should give a result equal to π_N:

$$\int D_{12...N}\, d\Omega_{12...N} = \pi_N. \tag{1}$$

In accordance with the regressive method, we have reduced, simple, double, etc., densities in phase corresponding to this situation:

$$\mu_{N^1} = N \int D_{12\dots N}\, d\Omega_{2\dots N},$$

$$\mu_{N^{12}} = N(N-1) \int D_{12\dots N}\, d\Omega_{34\dots N}.$$

$$\dots$$

In particular

$$\int \mu_{N^1}\, d\Omega_1 = N\pi_N.$$

A new regression relating to the occupation number leads us to new reduced, simple, double, etc., densities:

$$\mu_1 = \sum_N \mu_{N^1}, \tag{2}$$

$$\mu_{12} = \sum_N \mu_{N^{12}},$$

$$\dots$$

These last quantities have a fundamental part to play in our studies. We note the following relation:

$$\langle N \rangle = \int \mu_1\, d\Omega_1. \tag{3}$$

The densities in phase that we have just defined correspond in the usual way to spatial densities:

$$n_1 = \int \mu_1\, d^3 p_1,$$

$$n_{12} = \int \mu_{12}\, d^3 p_1\, d^3 p_2.$$

There is no relation which, as in the case when the occupation is definitely determined, will allow us to pass by regression from the double density in phase μ_{12} to the simple density μ_1 or from n_{12} to n_1. As a result situations where we should have

$$n_{12} = n_1 n_2 \quad \text{or} \quad \mu_{12} = \mu_1 \mu_2$$

are possible and the suspicion with which we surrounded them in section 1.3 no longer holds. This possibility cannot be stressed too much.

We shall hardly be speaking of μ_{N1}, μ_{N12}, etc., any more. From now on we can therefore reserve the term of reduced density for $\mu_1, \mu_{12}, \dots$

It will often be convenient to denote π_0 by D_0, and this quantity completes the list of occupation densities.

Generalizing formulae (2) we can finally write the general expression for the density $\mu_{12...M}$ as a function of the D:

$$\mu_{12...M} = \sum_{N \geq M} \frac{N!}{(N-M)!} \int D_{12...N} \, d\Omega_{M+1...N}. \tag{4}$$

Likewise the values of D can be expressed as a function of the values of μ. The following relation can be verified without difficulty:

$$D_{12...M} = \frac{1}{M!} \sum_{N=0}^{\infty} \frac{(-1)^N}{N!} \int \mu_{12...(M+N)} \, d\Omega_{M+1...(M+N)}. \tag{5}$$

The correspondence between the values of D and those of μ is therefore one-to-one. Bearing in mind the normalization condition for the values of π, formula (5) can be generalized even to the case of D_0:

$$D_0 = 1 - \int \mu_1 \, d\Omega_1 + \tfrac{1}{2} \int \mu_{12} \, d\Omega_{12} - \tfrac{1}{6} \int \mu_{123} \, d\Omega_{123} + \dots \tag{6}$$

2.3. Calculation of Averages

When the occupation is definitely known we have defined the average of a physical quantity F and we have denoted it by the symbol

$$\langle F \rangle_N. \tag{1}$$

The averages in which we are now interested can be obtained by weighting the earlier averages by the values of π_N:

$$\langle F \rangle = \sum_N \pi_N \langle F \rangle_N. \tag{2}$$

We have shown how the averages (1) could be expressed in all the cases of interest in terms of the reduced densities. In the case of the more general averages (2) the expressions are changed simply by substituting the new reduced densities for the old ones. This result comes from the linearity of the formulae with respect to the old reduced densities.

In making the same change the mean square of the fluctuation of a quantity does not alter its form either. Formula (4) of section 1.4 is therefore still correct and can be applied even when the partial volume is the same as the total volume.

2.4. Evolution

In a system which neither gains nor loses particles the π_N are constants, and the densities in phase and of occupation each continue to satisfy the corresponding Liouville equation. As a result the chain of recurrence equations of motion can be applied to our new reduced densities.

It should be noted that the method of reduction which allowed us to pass from equation (3) of section 1.9 to equation (2) of section 1.9 no longer applies here. Reduction of an equation in the chain no longer leads to the equation which precedes it in logical order. Equations of a new type are thus obtained.

The case of the diffusion equations (diffusion of the particles, of the momentum, of the energy) is a little more complicated, since local averages and local fluctuation averages appear. In fact, it can be shown that if the new reduced densities are substituted for the old ones in the old definitions of the local quantities, the formalism here is still unchanged.

Finally, everything can be expressed in terms of the reduced densities; physical calculations will be made essentially with the first reduced, simple, double, or triple densities. These few quantities which only provide us with information on the fluid point by point do not tell us the exact distribution of the values of π_N, even though earlier they had led us to abandon our interest in the system's trajectory in phase space.

2.5. Analysis of Correlations

To analyse† these correlations conveniently it is appropriate to introduce a new chain of phase functions. We put

$$
\begin{aligned}
\mu_1 &= \Gamma^{-1} \, e^{\varphi_1}, \\
\mu_{12} &= \Gamma^{-2} \, e^{\varphi_1 + \varphi_2 + \varphi_{12}}, \\
\mu_{123} &= \Gamma^{-3} \, e^{\varphi_1 + \varphi_2 + \varphi_3 + \varphi_{12} + \varphi_{13} + \varphi_{23} + \varphi_{123}},
\end{aligned}
\tag{1}
$$

and so on. The parameter Γ is introduced here for the new functions φ to be absolutely dimensionless, which is not the case for μ.

---

† We owe this method to H. S. Green who, nevertheless, applied it to the case when the occupation is definitely known. Its extension to the case when it is not was used by Nettleton and Green (1958).

Occupation

It is convenient to put

$$\Gamma = h^3, \tag{2}$$

h being Planck's constant which has the desired dimensions. As the μ are positive or zero, the φ are defined unambiguously. They are real quantities which can be either positive or negative.

A first approximation boils down to assuming that all the φ are zero except the $\varphi_1, \varphi_2, \ldots$ which depend only on the phase of a single particle. In future we shall call this the perfect fluid approximation. A better approximation consists of assuming that all the φ are zero except for the φ with only one suffix and the φ with two suffices, $\varphi_{12}, \varphi_{13}$, etc. This is Kirkwood's approximation. The statistics are then entirely defined by μ_1 and μ_{12}. In particular

$$\mu_{123} = \frac{\mu_{12}\mu_{13}\mu_{23}}{\mu_1\mu_2\mu_3}. \tag{3}$$

The following approximation retains the φ_{123} but neglects the φ_{1234} and similar higher-order functions. In a general manner systematic introduction of the functions φ permits rigorous solutions to be carried out by successive approximation. It is also a valuable guide for approximate solutions.

Analysis of the chain of the $D_{12\ldots N}$ can be carried out similarly, with a few slight differences. We put

$$D_1 \quad = \frac{1}{1!} \Gamma^{-1} D_0 e^{h_1},$$

$$D_{12} \quad = \frac{1}{2!} \Gamma^{-2} D_0 e^{h_1+h_2+h_{12}},$$

$$D_{123} = \frac{1}{3!} \Gamma^{-3} D_0 e^{h_1+h_2+h_3+h_{12}+h_{13}+h_{23}+h_{123}}. \tag{4}$$

The introduction here of the factorials and of D_0 will be justified later. It is not necessary to press the matter here since the occupation densities do not have the basic statistical character of the reduced densities. The functions $h_1, h_{12}, \ldots$ have a simple meaning in thermodynamic equilibrium.

The ideas in this section will not be applied all at once: the method described in the next section will help to prepare the ground.

40

2.6. The Cumulant Method

The relations between the values of μ and the values of D imply the use of series that in general hardly converge at all since the dominant terms occur after a number of terms which may be of the order of magnitude of the Avogadro number. This inconvenience appears clearly in calculating the average occupation number, but it is a general one.

A first step that can be taken to reduce this inconvenience is supplied by the cumulant method. Before concerning ourselves with the continuous case, which is the one which is of interest to us here, we shall first deal with a discrete problem.

We shall consider a discrete series of suffices $i, j, k, \ldots$ which will generally be infinite and a series of numbers

$$C_i, C_j, \ldots, C_k, \ldots,$$

then another series of numbers characterized by two suffices

$$C_{ii}, C_{ij}, C_{ik}, \ldots, C_{jj}, C_{jk}, \ldots,$$

then another series characterized by three suffices

$$C_{ijk},$$

which can still take on all possible values, and so on. In order to obtain the maximum generality no symmetry features are assigned *a priori* to the C.

With the help of this first collection we shall form a new collection of numbers of the same type $A_i, A_{ij}, A_{ijk}, A_{ijkl}$ as follows:

$$
\begin{aligned}
A_i &= C_i, \\
A_{ij} &= C_{ij} + C_i C_j, \\
A_{ijk} &= C_{ijk} + C_{ij}C_k + C_{ik}C_j + C_{jk}C_i + C_i C_j C_k, \\
A_{ijkl} &= C_{ijkl} + C_{ijk}C_l + C_{ijl}C_k + C_{ikl}C_j + C_{jkl}C_i \\
&\quad + C_{ij}C_{kl} + C_{ik}C_{jl} + C_{il}C_{jk} + C_{ij}C_k C_l \\
&\quad + C_{ik}C_j C_l + C_{il}C_j C_k + C_{jk}C_i C_l \\
&\quad + C_{jl}C_i C_k + C_{kl}C_i C_j + C_i C_j C_k C_l.
\end{aligned}
\tag{1}
$$

The rule for the formation of A is obvious. In effective calculations of any higher-rank A no term must be forgotten and we must know how to evaluate the number of terms of each type.

The definitions above have been given in the case of distinct suffices. When certain suffices are the same the desired expressions are obtained by identifying these suffices in the original expressions. For example,

$$A_{iii} = C_{iii} + 3C_{ii}C_i + C_i^3.$$

The C can be expressed in terms of the A:

$$C_i = A_i,$$
$$C_{ij} = A_{ij} - A_i A_j,$$
$$C_{ijk} = A_{ijk} - A_{ij}A_k - A_{ik}A_j - A_{jk}A_i + 2A_i A_j A_k,$$
$$C_{ijkl} = A_{ijkl} - A_{ijk}A_l - A_{ijl}A_k - A_{ikl}A_j - A_{jkl}A_i$$
$$- A_{ij}A_{kl} - A_{ik}A_{jl} - A_{il}A_{jk}$$
$$+ 2A_{ij}A_k A_l + 2A_{ik}A_j A_l + 2A_{il}A_j A_k$$
$$+ 2A_{jk}A_i A_l + 2A_{jl}A_i A_k + 2A_{kl}A_i A_j$$
$$- 6A_i A_j A_k A_l,$$
$$\cdots \tag{2}$$

These formulae resemble the preceding ones except that the signs now alternate and numerical factors appear which are successively equal to 2!, 3!, ...

The numbers C are called the cumulants of the numbers A. A simple property of the cumulants is that all the cumulants with more than one suffix are zero if the $A_{ij}, A_{ijk} \ldots$ are products:

$$A_{ij} = A_i A_j,$$
$$A_{ijk} = A_i A_j A_k,$$
$$\cdots$$

Let us now introduce a series of auxiliary parameters $z_i, z_j, z_k, \ldots$ We consider the two functions

$$F(z) = 1 + \sum_i A_i z_i + \frac{1}{2!} \sum_{i,j} A_{ij} z_i z_j + \frac{1}{3!} \sum_{i,j,k} A_{ijk} z_i z_j z_k + \cdots \tag{3}$$

$$G(z) = \sum_i C_i z_i + \frac{1}{2!} \sum_{i,j} C_{ij} z_i z_j + \frac{1}{3!} \sum_{i,j,k} C_{ijk} z_i z_j z_k + \cdots . \tag{4}$$

In the above sums each suffix takes up every possible value. These two functions are connected by the relation

$$F = \exp G. \tag{5}$$

The formalism can be generalized when the suffices form a continuous series. We now replace the discrete suffix i by the continuous suffix xp. We can then write the cumulants of our reduced densities:

$$\mu_1,$$

$$\mu_{12} - \mu_1\mu_2,$$

$$\mu_{123} - \mu_{12}\mu_3 - \mu_{13}\mu_2 - \mu_{23}\mu_1 + 2\mu_1\mu_2\mu_3,$$

$$\cdots \tag{6}$$

In the case of a perfect fluid all the cumulants are zero except for the first one.

The auxiliary parameters z_i are replaced by an auxiliary function $f(x, p)$ and the functions F and G by functionals of f:

$$F = 1 + \int \mu_1 f_1 \, d\Omega_1 + \frac{1}{2!} \int \mu_{12} f_1 f_2 \, d\Omega_{12}$$

$$+ \frac{1}{3!} \int \mu_{123} f_1 f_2 f_3 \, d\Omega_{123} + \cdots \tag{7}$$

$$G = \int \mu_1 f_1 \, d\Omega_1 + \frac{1}{2!} \int (\mu_{12} - \mu_1\mu_2) f_1 f_1 \, d\Omega_{12}$$

$$+ \frac{1}{3!} \int (\mu_{123} - \mu_{12}\mu_3 - \mu_{13}\mu_2 - \mu_{23}\mu_1 + 2\mu_1\mu_2\mu_3)$$

$$\times f_1 f_2 f_3 \, d\Omega_{123} + \cdots. \tag{8}$$

The relation (5) still holds.

A first application of the formulae (6), (7) and (8) is to obtain an exponential expression for D_0. We take $f = -1$. In accordance with formula 2 (6) we obtain

$$D_0 = \exp\left[-\int \mu_1 \, d\Omega_1 + \frac{1}{2}\int (\mu_{12} - \mu_1\mu_2) \, d\Omega_{12}\right.$$

$$- \frac{1}{6}\int (\mu_{123} - \mu_{12}\mu_3 - \mu_{13}\mu_2 - \mu_{23}\mu_1 + 2\mu_1\mu_2\mu_3)$$

$$\left. \times \, d\Omega_{123} + \cdots\right]. \tag{9}$$

A variational calculation now allows us to obtain the other densities in phase and of occupation. Let us apply to f an infinitesimal variation δf. Bearing in mind the symmetry we obtain

$$\delta F = \int \mu_1 \, \delta f_1 \, d\Omega_1 + \int \mu_{12} \, \delta f_1 f_2 \, d\Omega_{12}$$

$$+ \frac{1}{2}\int \mu_{123} \, \delta f_1 f_2 f_3 \, d\Omega_{123} + \cdots, \tag{10}$$

43

Occupation

$$\delta \exp G = \left\{ \int \mu_1 \, \delta f_1 \, \mathrm{d}\Omega_1 + \int (\mu_{12} - \mu_1\mu_2) \, \delta f_1 f_2 \, \mathrm{d}\Omega_{12} \right.$$
$$+ \tfrac{1}{2} \int (\mu_{123} - \mu_{12}\mu_3 - \mu_{13}\mu_2 - \mu_{23}\mu_1 + 2\mu_1\mu_2\mu_3)$$
$$\left. \times f_2 f_3 \, \delta f_1 \, \mathrm{d}\Omega_{123} + \cdots \right\} \exp G. \tag{11}$$

By making in the integrand the factor of δf_1, which is arbitrary, equal on the two sides, and by making f itself tend towards -1, we obtain in accordance with equation (5) a new expression for D_1:

$$D_1 = \left\{ \mu_1 - \int (\mu_{12} - \mu_1\mu_2) \, \mathrm{d}\Omega_2 + \tfrac{1}{2} \int (\mu_{123} - \mu_{12}\mu_3 - \mu_{13}\mu_2 \right.$$
$$\left. - \mu_{23}\mu_1 + 2\mu_1\mu_2\mu_3) \, \mathrm{d}\Omega_{23} + \cdots \right\} D_0. \tag{12}$$

The same technique applied to the second variation gives

$$D_{12} = \tfrac{1}{2} \left\{ \mu_{12} - \int (\mu_{123} - \mu_{12}\mu_3) \, \mathrm{d}\Omega_3 + \tfrac{1}{2} \int (\mu_{1234} - \mu_{123}\mu_4 \right.$$
$$\left. - \mu_{124}\mu_3 - \mu_{12}\mu_{34} + 2\mu_{12}\mu_3\mu_4) \, \mathrm{d}\Omega_{34} + \cdots \right\} D_0, \tag{13}$$

and more generally

$$D_{12\ldots N} = \frac{1}{N!} \left\{ \mu_{12\ldots N} - \int (\mu_{12\ldots N+1} - \mu_{12\ldots N} \, \mu_{N+1}) \, \mathrm{d}\Omega_{N+1} \right.$$
$$+ \frac{1}{2} \int (\mu_{12\ldots N+1,N+2} - \mu_{12\ldots N+1} \, \mu_{N+2} - \mu_{12\ldots N,N+1} \, \mu_{N+1}$$
$$\left. - \mu_{12\ldots N} \, \mu_{N+1,N+2} + 2\mu_{12\ldots N} \, \mu_{N+1} \, \mu_{N+2}) \, \mathrm{d}\Omega_{N+1,N+2} + \cdots \right\} D_0. \tag{14}$$

In the expression for D_{12} we see cumulants of a new type appear where the pair of suffices 1 and 2 plays a particular part. A similar situation occurs for higher-order densities.

2.7. Perfect Fluid

With our last definition of the perfect fluid the reduced densities are "factorized":

$$\mu_{12\ldots N} = \mu_1\mu_2 \cdots \mu_N. \tag{1}$$

The reduced densities have no correlation. The same is true of the space densities

$$n_{12} = n_1 n_2.$$

44

Since the cumulants figuring in the developments of the last section are zero for the most part, the densities in occupation also have simple expressions:

$$D_0 = \exp(-\langle N \rangle),$$
$$D_1 = D_0 \mu_1,$$
$$D_{12} = \frac{1}{2!} D_0 \mu_1 \mu_2. \tag{2}$$

From this we can deduce the chain of occupation probabilities:

$$\pi_M = \frac{1}{M!} (\langle N \rangle)^M \exp(-\langle N \rangle) \tag{3}$$

which is correctly normalized. The π_M chain first increases with M particularly if the average occupation number is high, then decreases because of the effect of the factorial. To get an idea of the behaviour of π_M in the vicinity of the maximum we can use Stirling's formula, which is valid for large values of M:

$$\ln M! = M \ln M - M + \tfrac{1}{2} \ln M + \tfrac{1}{2} \ln 2\pi + \cdots, \tag{4}$$

from which we obtain an approximate expression for π_M:

$$\pi_M \sim (2\pi M)^{-1/2} \exp\left(M \ln \frac{\langle N \rangle}{M} + M - \langle N \rangle \right). \tag{5}$$

This function of M is maximal for M close to $\langle N \rangle$ and has the appearance of a Gaussian distribution in the vicinity of this maximum. In this region we can write

$$\pi_M = (2\pi \langle N \rangle)^{-1/2} \exp\left(-\frac{(M - \langle N \rangle)^2}{2\langle N \rangle} \right). \tag{6}$$

In the case of the perfect fluid the $D_{12\ldots N}$ are therefore factorized at the same time as the μ, which was not rigorously true when the number of particles was definitely known. Here, in the same way as the φ are reduced to φ_1, the h are reduced to h_1. In addition we note that the simplicity of the result justifies the introduction of the factorials in the definition of h.

From the point of view of evolution a fluid assumed to be perfect does not keep this property in the course of time. Nevertheless an approximation which is useful in certain cases can be found here. Let us, in fact, introduce this approximation into the first two recurrence equations of motion. We introduce the self-consistent

field:
$$F_1'' = \int F_{12} n_2 \, d^3 r_2. \tag{7}$$

These equations can be written respectively as

$$\left[\frac{\partial}{\partial t} + \frac{p_1}{m} \frac{\partial}{\partial x_1} + (X_1 + X_1'') \frac{\partial}{\partial p_1} \right] \mu_1 = 0, \tag{8}$$

$$\left[\frac{\partial}{\partial t} + \frac{p_1}{m} \frac{\partial}{\partial x_1} + (X_1 + X_1'') \frac{\partial}{\partial p_1} + \frac{p_2}{m} \frac{\partial}{\partial x_2} \right.$$
$$\left. + (X_2 + X_2'') \frac{\partial}{\partial p_2} + X_{12} \left(\frac{\partial}{\partial p_1} - \frac{\partial}{\partial p_2} \right) \right] \mu_1 \mu_2 = 0. \tag{9}$$

The sum of the applied field and the self-consistent field appears in equation (8). This equation is convenient for studying the evolution of μ_1. The next equation has no solution because of the presence of the intermolecular forces. If, nevertheless, we neglect this, this equation is a simple consequence of the preceding one. This kind of approximation is suitable for the study of strongly ionized gases: in these media the Coulomb forces between the particles are predominant. At least in studying transitory phenomena it is more convenient to study their influence in the synthetic form of the self-consistent field than by their binary effects, so that we may take only the self-consistent effect into consideration. In the case of short-range forces this approximation ceases to be of interest, however.

Equation (8) is known as the Vlasov equation.

2.8. Molecular Disorder

The set of relations which we have established with the help of the cumulants is general, but the advantage of these relations appears only if the correlations have a particular structure—a structure of which the perfect fluid is a prime example.

We shall say that the system is in a state of molecular disorder if the correlations are zero between regions which are far apart—the perfect fluid corresponds to the case when they are zero throughout.

Let us assume the system's occupation number to be very high. Let us also define at each point a characteristic distance

$$d_1 = n_1^{-1/3}.$$

In a cube with edge length d_1 centred on the point in question there is on the average only a single particle. This statement can not be taken literally, since it is rigorous only in a region where the fluid would be uniform, but our considerations are qualitative. If the system contains a large number of particles the characteristic distance is extremely small when compared with the dimensions of the container. A distance which is a hundred or a thousand times the characteristic distance is large from the microscopic point of view and small from the macroscopic point of view.

The correlations can be analysed by considering the quantities $\varphi_{12}, \varphi_{123}, \varphi_{1234}, \cdots$

In a state of molecular disorder around each point in space there is a correlation sphere—a sphere whose radius is small from the macroscopic point of view. Every φ is zero if the correlation spheres centred respectively on each of the particles concerned have no point common to all.

A state of molecular disorder shows integrals like

$$\int (n_{12} - n_1 n_2) \, d^3 r_2, \quad \int (\mu_{12} - \mu_1 \mu_2) \, d\Omega_2$$

in a new light. In principle, integration should be carried out over the whole of the available volume. In fact the integrand is zero almost everywhere, and the value of the integral is calculated over a very small region. The same is true of all the integrals which involve cumulants.

To show this effect of the cumulants which come into our calculations we put

$$a_{12} = e^{\varphi_{12}}, \quad a_{123} = e^{\varphi_{123}}, \quad a_{1234} = e^{\varphi_{1234}}, \quad \text{etc.}$$

The a are equal to unity when the corresponding φ is zero. We shall then use the following quantities:

$$\varepsilon_{12} = a_{12} - 1, \quad \varepsilon_{123} = a_{123} - 1,$$
$$\varepsilon_{1234} = a_{1234} - 1,$$

which are zero when the corresponding φ is zero. These two sets of quantities somewhat duplicate each other but their simultaneous use makes the notation easier.

Using our new notation we can, for example, rewrite the expression for D_0 as a function of the μ.

1. The direct expression:

$$D_0 = 1 - \int \mu_1 \, d\Omega_1 + \tfrac{1}{2} \int \mu_1 \mu_2 a_{12} \, d\Omega_2$$
$$- \tfrac{1}{6} \int \mu_1 \mu_2 \mu_3 a_{12} a_{13} a_{23} a_{123} \, d\Omega_{123} + \cdots;$$

2. The exponential expression:

$$D_0 = \exp\left\{- \int \mu_1 \, d\Omega_1 + \tfrac{1}{2} \int \mu_1\mu_2\varepsilon_{12} \, d\Omega_{12} \right.$$
$$- \tfrac{1}{6} \int \mu_1\mu_2\mu_3(a_{12}a_{13}a_{23}\varepsilon_{123} + \varepsilon_{12}\varepsilon_{13}\varepsilon_{23}$$
$$\left. + \varepsilon_{12}\varepsilon_{13} + \varepsilon_{12}\varepsilon_{23} + \varepsilon_{13}\varepsilon_{23}) \, d\Omega_{123} - \cdots \right\}.$$

For each term in the exponential all the integrations are spatially limited to a narrow region in the vicinity of the point 1 because of the effect of the ε, except of course for the last one. In the direct expression the term of rank M is of order of magnitude

$$\frac{1}{M!} \langle N \rangle^M,$$

whilst in the exponential expression the term of the same rank is of order of magnitude—assuming in addition that the values of a have an upper limit which is not much larger than unity—

$$\frac{1}{M!} \langle N \rangle \, P^{M-1},$$

where P is the average number of particles which may be in a sphere whose radius is the diameter of the correlation sphere.

These remarks are superficial. They are valid only for the first terms in each series. Let us consider in particular the series which figures in the exponential. We have taken no account of the number of terms figuring in each integrand. In addition, a term such as $\varepsilon_{12}\varepsilon_{23}$ or $\varepsilon_{13}\varepsilon_{23}$ has a "range" which is not the correlation diameter, but is twice as great. This term is the forerunner in higher-order integrands of "chains" which may be long enough to sweep the whole of the available volume. It will be seen, on the other hand, that a quantity such as $\varepsilon_{12}\varepsilon_{13}\varepsilon_{23}$ corresponds to a closed chain and to terms of a less all-embracing type.

All we need of our qualitative remarks is that the first terms in the exponential expansion are much more significant than the first terms of the direct expansion, provided that molecular disorder obtains.

Assuming that molecular disorder obtains at a given point in time, there is nothing to prove that it is maintained indefinitely. Long-range correlations could become established little by little. This is a difficult problem. The theory of gas dynamics, i.e. the dynamics of low-density media, is based on the hypothesis of molecular chaos. The idea of molecular chaos has points in common with the idea of molecular disorder, but it is more accurate and more particular. A comparative examination of them would be premature.

We have considered explicitly only fluids. It should not be thought that the situation is all that different in a solid. The existence of crystal lattices corresponds to the fact that μ_1, n_1 are then strongly structured. The characteristics of $a_{12}, a_{123}, \ldots$ which we have just examined are not necessarily modified all that much. It is inappropriate to dwell on this much here since, except for special cases, the classical description of solids is unsatisfactory.

2.9. Pressure in a Plasma

Intermolecular pressure, in the case of short-range forces, was defined clearly in the first chapter. The case of plasmas was left in abeyance.

Let us note first of all that, in a general way, the problem of pressure, or rather of pressures, in a plasma is complex since we must consider:

the pressure due to the particles, whether kinetic or intermolecular;

the magnetic pressure;

the radiation pressure—as in all media—but it may be more acute here because of the great aptitude of electrons for radiating.

We shall content ourselves here with a preliminary remark on intermolecular pressure—which would, moreover, be better called "interelectronic". We are dealing with a simplified plasma: it comprises only electrons which move in a continuous positive background which does not contribute to the pressure.

In section 6 of Chapter 1 we established once and for all the following relations:

$$\int n_{12}X_{12} = -\frac{\partial}{\partial y_1} p_{1xy}, \tag{1}$$

$$p_{Mxy} = -\frac{1}{2} \int_{4\pi} \int_0^\infty \int_0^\infty n_{JK}(x_K - x_J)(y_K - y_J)\frac{dW}{dr_{JK}}dr_{MJ}\,dr_{MK}\,d^2O. \tag{2}$$

49

These are general relations which are valid whatever the range of the forces. The quantity p_M has been given the name of pressure: on the other hand this terminology is not general and is appropriate only for the case of short-range forces.

In the case of a system of electrons, since one cannot count on the force to cut off the integration at short distances, we must look for this cut-off to the statistics. We must first of all assume that the situation is that of molecular disorder. We then start by introducing the self-consistent field:

$$\int n_{12}F_{12} \, \mathrm{d}^3r_2 = n_1 \int n_2 F_{12} \, \mathrm{d}^3r_2 + \int (n_{12} - n_1 n_2) \, F_{12} \, \mathrm{d}^3r_2$$

$$= n_1 F_1'' + \int (n_{12} - n_1 n_2) \, F_{12} \, \mathrm{d}^3r_2 . \tag{3}$$

Only the last integral will be reduced to the divergence of a tensor. We write

$$\pi_{Mxy} = -\frac{1}{2} \int\limits_{4\pi} \int\limits_{0}^{\infty} \int\limits_{0}^{\infty} (n_{JK} - n_J n_K)(x_K - x_J)(y_K - y_J)$$

$$\times \frac{\mathrm{d}W}{\mathrm{d}r_{JK}} \, \mathrm{d}r_{MJ} \, \mathrm{d}r_{MK} \, \mathrm{d}^2 O . \tag{4}$$

The analysis of the forces we gave in Chapter 1 can be generalized here. We see at once that we can now write an equation that is similar to (1) as follows:

$$\int (n_{12} - n_1 n_2) \, X_{12} \, \mathrm{d}^3r_2 = -\frac{\partial}{\partial y_1} \pi_{1xy} . \tag{5}$$

We shall say that π_{xy} is the interelectronic pressure without trying to hide the fact that this may be an abuse of language. This pressure has not at all got the structure of the intermolecular pressure. When the forces are not purely Coulomb forces, when, for example, a hard core is ascribed to charged particles, the above formalism does not go deep enough. We shall return to this when we discuss mixtures.

2.10. Mixtures

We shall now consider a system made up of two kinds of molecule. The simplest example is a mixture of two gases. The problem is in no way a chemical one and we assume in fact that the two constituents do not react with each other.

These two constituents are called a and b respectively. We shall now have to manipulate a larger number of quantities:

masses $\qquad\qquad\qquad\quad m_a, m_b$

applied fields $\qquad\qquad\quad \boldsymbol{F}_{a1}, \boldsymbol{F}_{b1}$

intermolecular forces $\quad \boldsymbol{F}_{a12}, \boldsymbol{F}_{b12}, \boldsymbol{F}_{ab12}, \boldsymbol{F}_{ba12}$

simple densities $\qquad\quad n_{a1}, n_{b1}$

$\qquad\qquad\qquad\qquad\quad \mu_{a1}, \mu_{b1}$

double densities $\qquad\quad \mu_{a12}, \mu_{b12}, \mu_{ab12}, \mu_{ba12}.$

By virtue of the principle of action and reaction the two mixed forces are equal:

$$\boldsymbol{F}_{ab12} = \boldsymbol{F}_{ba12}.$$

We shall denote the common value by $\boldsymbol{F}_{12}$. A similar relation for the double densities in phase

$$\mu_{ab12} = \mu_{ba12}$$

is in general not valid. In particular, if there is no correlation between the two fluids this would mean that

$$\mu_{a1}\mu_{b2} = \mu_{b1}\mu_{a2},$$

which cannot be exact except in special cases.

The two equations of evolution for the simple density in phase are respectively

$$\left. \begin{aligned}
&\left(m_a \frac{\partial}{\partial t} + m_a u_1 \frac{\partial}{\partial x_1} + X_{a1} \frac{\partial}{\partial u_1}\right)\mu_{a1} \\
&\quad + \frac{\partial}{\partial u_1} \int (X_{a12}\mu_{a12} + X_{12}\mu_{ab12})\, \mathrm{d}\Omega_2 = 0, \\
&\left(m_b \frac{\partial}{\partial t} + m_b u_1 \frac{\partial}{\partial x_1} + X_{b1} \frac{\partial}{\partial u_1}\right)\mu_{b1} \\
&\quad + \frac{\partial}{\partial u_1} \int (X_{b12}\mu_{b12} + X_{12}\mu_{ba12})\, \mathrm{d}\Omega_2 = 0.
\end{aligned} \right\} \quad (1)$$

From this we can deduce for each fluid the equations of continuity and the equations of motion:

$$\frac{\partial}{\partial t} n_{a1} + \frac{\partial}{\partial x_1} n_{a1} \langle u_1 \rangle_a = 0,$$

$$\frac{\partial}{\partial t} n_{b1} + \frac{\partial}{\partial x_1} n_{b1} \langle u_1 \rangle_b = 0. \qquad (2)$$

51

$$m_a\left(\frac{\partial}{\partial t}\, n_{a1}\, \langle u_1\rangle_a + \frac{\partial}{\partial y_1}\, n_{a1}\, \langle u_1 v_1\rangle_a\right) = n_{a1} X_{a1}$$

$$+ \int (X_{a12} n_{a12} + X_{12} n_{ab12})\, \mathrm{d}^2 r_2,$$

$$m_b\left(\frac{\partial}{\partial t}\, n_{b1}\, \langle u_1\rangle_b + \cdots\right) = n_{b1} X_{b1} + \int (X_{b12} n_{b12} + X_{12} n_{ba12})\, \mathrm{d}^3 r.$$

$$(3)$$

It often occurs that the two fluids remain solidary: they have essentially the same average velocity. There is therefore motion of the whole accompanied by a slight relative motion: the latter is the phenomenon of diffusion. The movement of the whole is above all controlled by the effects of inertia. It is therefore natural to introduce the mass density:

$$\varrho_1 = m_a n_{a1} + m_b n_{b1}, \tag{4}$$

and to define an average velocity which will play a major part by a weighting of the masses similar to the one above:

$$\varrho_1 \langle c_1\rangle = m_a n_{a1} \langle c_1\rangle_a + m_b n_{b1} \langle c_1\rangle_b. \tag{5}$$

This velocity $\langle c_1\rangle$ is the velocity of convection. From equations (2) we can deduce the equation for the conservation of mass:

$$\frac{\partial}{\partial t}\varrho + \frac{\partial}{\partial x}\varrho \langle u\rangle = 0, \tag{6}$$

and the convection equation from equations (3):

$$\varrho\left[\frac{\partial}{\partial t} + \langle v_1\rangle \frac{\partial}{\partial y_1}\right]\langle u_1\rangle + \frac{\partial}{\partial y_1} k_{1xy} = n_{a1} X_{a1} + n_{b1} X_{b1}$$

$$+ \int \{X_{a12} n_{a12} + X_{12}(n_{ab12} + n_{ba12}) + X_{b12} n_{b12}\}\, \mathrm{d}^3 r_2. \tag{7}$$

The symmetrical kinematic pressure tensor can be introduced as follows:

$$k_{1xy} = m_a n_{a1} \langle u_1 v_1\rangle_a + m_b n_{b1} \langle u_1 v_1\rangle_b - \varrho \langle u_1\rangle \langle v_1\rangle. \tag{8}$$

We assume that the forces are short-range ones. It is then interesting to introduce as well the intermolecular pressure. The arguments in section 1.6 can be applied without special mention of the two integrals

$$\int F_{a12} n_{a12}\, \mathrm{d}^3 r_2 \quad \text{and} \quad \int F_{b12} n_{b12}\, \mathrm{d}^3 r_2.$$

The same is not true example of the integral

$$\int F_{12} n_{ab12}\, \mathrm{d}^3 r_2.$$

Let us return to the arguments of section 1.6. One of the essential points was that the sum of the forces exerted on each other by molecules of the same kind which are all contained in a closed volume is zero. Here we must consider the sum of the forces exerted by the molecules of kind b on those of kind a: this sum is not zero. On the other hand, the argument can be generalized for the sum of the two integrals:

$$\int F_{12}(n_{ab12} + n_{ba12})\, \mathrm{d}^3r_2$$

and it is this which allows us to introduce an intermolecular pressure p_{xy} by a formula which generalizes equation (16) of section 1.6. The average force derives from this tensor:

$$\int \{X_{a12}n_{a12} + X_{12}(n_{ab12} + n_{ba12}) + X_{b12}n_{ab12}\}\, \mathrm{d}^3r_2 = -\frac{\partial}{\partial y_1} p_{1xy}.$$
(9)

We now define the total pressure

$$P_{xy} = k_{xy} + p_{xy},$$

which allows us to put the convection equation in the same form as the hydrodynamic equation:

$$\varrho \left[\frac{\partial}{\partial t} + \langle v \rangle \frac{\partial}{\partial y}\right] \langle u \rangle + \frac{\partial}{\partial y} P_{xy} = n_{a1}X_{a1} + n_{b1}X_{b1}. \quad (10)$$

In writing down equations (6) and (10) we have not exhausted the contents of equations (2) and (3). In particular, we still have to write down a diffusion equation similar to the convection equation. We shall not write it down in general because the conventions which are most suitable for describing the relative motion include a certain arbitrary factor which allows us to adapt the writing to the particular problem being treated: mixtures of neutral particles, of charged particles and neutral particles, or of charged particles of opposite signs.

But whatever the choice adopted we shall not succeed in arranging the intermolecular actions so as to make them derivable from a tensor similar to the pressure tensor. This feature is essential in the theory of electric conductivity, the diffusion current in this case being the same as the conduction current. On the other hand, the convection current cannot be subjected to any resistive effect because the hydrodynamic formula leaves no room for a braking in simple proportion to this current. In this connexion there can only be friction of the viscous type conditioned by the gradients of the pressure components. We should add, since it is a known fact, that the case of Coulomb forces, which are long-range forces, requires some preparatory work.

2.11. Mixtures. The Case of a Plasma

We shall now consider a fluid medium containing a certain number of kinds a, b, c which, for the sake of simplification, we shall assume cannot be transformed into each other. Let e_a be the electric charge of kind a: it can if necessary be zero.

As in the last section, the mass density will be written as

$$\varrho = \sum_a m_a n_a, \tag{1}$$

and the velocity of convection is given by the formula

$$\varrho \langle c \rangle = \sum_a m_a n_a \langle c_a \rangle. \tag{2}$$

The electrical density is

$$\sigma = \sum_a e_a n_a. \tag{3}$$

The electric current can be expressed in the form

$$J = \sum_a e_a n_a \langle c_a \rangle = \sigma \langle c \rangle + \sum_a e_a n_a [\langle c_a \rangle - \langle c \rangle] \tag{4}$$

where the first term is the convection current and where the second term, j, is the diffusion current. This current exists only if there are several kinds, even if they are all neutral except one.

The conservation equations relating to each kind can be derived easily from the equation for the conservation of mass

$$\frac{\partial \varrho}{\partial t} + \frac{\partial}{\partial y} \varrho \langle u \rangle = 0 \tag{5}$$

and the equation for the conservation of charge

$$\frac{\partial \sigma}{\partial t} + \frac{\partial}{\partial y} J_y = 0. \tag{6}$$

We shall use

$$F_{ab12}$$

to denote the exact force exerted by the particle $b2$ on the particle $a1$ and

$$F_{ab12}^{C}$$

to denote the corresponding Coulomb force. The difference

$$F_{ab12} - F_{ab12}^{C}$$

is a short-range force. We write

$$C_{12x} = -\frac{\partial}{\partial x_1} \frac{1}{r_{12}},$$

which gives

$$F_{ab12}^{C} = e_a e_b C_{12}.$$

We also write

$$F'_{ab1} = \frac{1}{n_{a1}} \int F_{ab12} n_{ab12} - F_{ab12}^{C} n_{a1} n_{b2}) \, d^3 r_2, \tag{7}$$

$$F''_{ab1} = \int F_{ab12}^{C} n_{b2} \, d^3 r_2 = e_a e_b \int C_{12} n_{b2} \, d^3 r_2. \tag{8}$$

The quantity (7) represents what we shall call the inter-particle force field relative to the action of the kind b on the kind a—to generalize the definition of (5) of section 1.5 with the action of one kind on itself as a special case. The quantity (8) derives from the idea of collective forces. The collective electric field is by definition

$$E_1 = \sum_b e_b \int C_{12} n_{b2} \, d^3 r_2 = \int C_{12} \sigma_2 \, d^3 r_2. \tag{9}$$

We note that integral (7) is local in nature when molecular disorder obtains, as we shall assume.

The equation of motion of the kind a is as follows:

$$m_a \frac{\partial}{\partial t} \left[n_a \langle u_a \rangle + \frac{\partial}{\partial y} n_a \langle u_a v_a \rangle \right] = n_a \left(X_a + \sum_b X''_{ab} \right) + n_a \sum_b X'_{ab}. \tag{10}$$

From this we can derive the hydrodynamic equation and the equation describing the evolution of the diffusion current which are, respectively,

$$\varrho \left[\frac{\partial}{\partial t} + \langle v \rangle \frac{\partial}{\partial y} \right] \langle u \rangle$$

$$+ \frac{\partial}{\partial y} \left[\sum_a m_a n_a \langle u_a v_a \rangle - \varrho \langle u \rangle \langle v \rangle \right] = \sum_a n_a X_a + \sigma E_x + \sum_{a,b} n_a X'_{ab}, \tag{11}$$

$$\frac{\partial j_x}{\partial t} + \frac{\partial}{\partial y} j_x \subset v \supset + j_y \frac{\partial}{\partial y} \subset u \supset$$

$$+ \frac{\partial}{\partial y} \sum_a e_a n_a \subset ((u_a - \subset u \supset)(v_a - \subset v \supset) \supset$$

$$- \frac{\sigma}{\varrho} \frac{\partial}{\partial y} \sum_a m_a n_a \subset ((u_a - \subset u \supset)(v_a - \subset v \supset) \supset$$

$$- \sum_{a,b} \left(\frac{e_a}{m_a} - \frac{\sigma}{\varrho} \right) n_a X'_{ab} = \sum_a \left(\frac{e_a}{m_a} - \frac{\sigma}{\varrho} \right) n_a X_a$$

$$+ E_x \sum_a \left(\frac{e_a}{m_a} - \frac{\sigma}{\varrho} \right) e_a n_a. \tag{12}$$

The remarks of the last section relating to the hydrodynamic equation are still valid here. Equation (11) can be written in the form

$$\varrho \left(\frac{\partial}{\partial t} + \subset v \supset \frac{\partial}{\partial y} \right) \subset u \supset = \sum_a n_a X_a + \sigma E_x - \frac{\partial}{\partial y} P_{xy}, \tag{13}$$

which contains a pressure tensor.

For all its generality the equation of the diffusion current presents an awkward complication. Nevertheless it brings forth useful points. The first relates to microscopic actions between particles, i.e. the quantity

$$\sum_{a,b} \left(\frac{e_a}{m_a} - \frac{\sigma}{\varrho} \right) n_a F'_{ab}. \tag{14}$$

It does not possess the symmetry properties characterizing

$$\sum_{a,b} n_a F'_{ab}.$$

It is therefore impossible to define here a tensor of which it would systematically be the divergence.

Let us imagine, to crystallize our ideas, a situation which has the following characteristics:

(1) the hydrodynamic velocity is zero;
(2) the conditions are permanent;
(3) the situation is uniform;
(4) the only field that counts is the electric field;
(5) there is no space charge.

These hypotheses leave the field open for a permanent current j. Equation (12) then reduced to the following:

$$E \sum_a (e_a^2/m_a) n_a = - \sum_{ab} (e_a/m_a) n_a F'_{ab}. \qquad (15)$$

If we had been able systematically to reduce the second member to a divergence it would have been zero by virtue of the third hypothesis. We should then have been able to state the following law: in any uniform ionized medium through which a current is passing the electric field is zero. We should therefore have obtained a result which was awkward and contradicts the elementary properties of conductors and *Ohm's law* which states them.

As we have already seen in respect of the general laws relating to the hydrodynamics of a pure fluid, our equations (12) and (13) merely form a framework which has to be filled in, either by fuller analysis or by empirical considerations. We shall have to content ourselves with empiricism here.

We notice that when the situation is uniform from point to point the densities n_{a1} and n_{b1} are constants. In addition n_{ab12} depends only on relative coordinates $r_2 - r_1$. If this function is even, the integral (7) is zero. Ohm's law—or its generalizations—when it applies has as a necessary consequence that the correlations between the particles are not even functions of the relative coordinates: the correlations are polarized and naturally such that they oppose the passage of the current: the medium shows resistivity.

This kind of argument, which is a little long but after all quite simple, once the meaning of the general equations has been made clear, allows us to obtain a number of interesting results. In order to go into the matter more deeply we must associate the equations relating to energy diffusion with the equations of motion. The method is applicable not only to ionized media but also to other mixtures.

CHAPTER 3

Statistical Entropy

3.1. Decoupling of a System

It is often useful to consider several parts in a system: we say for example that system A is comprised of two parts B and C.

The simplest case is the one where the two systems are not related to each other, each being held in a closed container—closed to particles and energy—and the two containers in question are located at some distance from each other. It is quite obvious that this kind of system holds no interest except insofar as we intend to bring the two systems into contact at a certain point in time.

In a more closely coupled situation the two systems are simply separated by a wall: this wall can be imagined as an impermeable metal foil. To be accurate, this foil will have to be incorporated into one of the two partial systems, the boundary being defined by the actual surface of the foil in contact with the other system. The foil in question is to some extent an ideal one which is subject to neither corrosion nor infiltration.

Another case is that of particles of different kinds enclosed in the same container. Particles of one kind will be included in system B and the particles of the other kind in system C.

From the more abstract point of view of describing systems in extension in phase it should be possible to give each partial system an independent representation. Each is described using a set of canonically conjugate variables and a Hamiltonian. There is thus a certain phase space corresponding to each of the systems B and C. The Hamiltonian H_A of the system A is made up by the sum of the functions H_B and H_C relating respectively to B and C to which we add an interaction term:

$$H_A = H_B + H_C + H_I.$$

It is indispensable that the interaction should not modify the canonical variables; in other words, the canonical variables of A should be obtained by adding the canonical variables of B considered by themselves and the canonical variables of C considered by themselves. The phase space of A is therefore the product of the phase space of B and the phase space of C, which is expressed by the relation:

$$d\Omega_A = d\Omega_B \, d\Omega_C. \tag{1}$$

The density in phase of A can be written as

$$D = D(P_B, P_C), \tag{2}$$

P_B being a point in the phase space of B and P_C being a point in the phase space of C. In general this density in phase will also be dependent on the time.

Starting with the density in phase of A we define "reduced" densities, one of which suffices to describe all that we may wish to know about B and the other all that we may wish to know about C:

$$D_B(P_B) = \int D(P_B, P_C) \, d\Omega_C,$$
$$D_C(P_C) = \int D(P_B, P_C) \, d\Omega_B. \tag{3}$$

The fact that the density is normalized means that the reduced densities are normalized. For example,

$$\int D_B(P_B) \, d\Omega_B = 1. \tag{4}$$

We say that the two systems have no correlations if the following relation is satisfied:

$$D = D_B D_C. \tag{5}$$

According to circumstances this kind of situation is accidental or permanent. It is clearly permanent if the interaction H_I is zero. It should be noted that the absence of interaction does not *ipso facto* imply the absence of correlations. The function $D(P_B, P_C)$ is subject only to the general mathematical conditions which are the rule for densities in phase.

3.2. Definition of Entropy

Entropy is a quantity whose definition comes from thermodynamics.

Statistical Entropy

Statistical considerations now lead us to introduce a quantity which will bear the same name.

This confusion of names is only partly justified: on the one hand it is possible to establish the identity of thermodynamic entropy and statistical entropy only to the extent that an idea of experimental origin and an idea of theoretical origin can be identified. In actual fact, thermodynamic entropy can be determined without referring to the molecular interpretation of macroscopic phenomena, whilst statistical entropy is essentially based on this interpretation, which limits its generality a little since the molecular description always involves a certain number of formal features. On the other hand, the justification that will be mentioned cannot be extended to variable conditions. However, we are introducing the definition of statistical entropy abruptly and will justify it later by use. Nevertheless we should say straightaway, to set the reader right, that statistical entropy provides an overall measure of the existence or absence of correlations between the various parts of a system.

Consider therefore a system, isolated or not, which can be in the situation of system A mentioned in section 1 but which can also be in the situation of system B mentioned in the same section. There is a complete and independent set of canonically conjugate variables. The entropy S is defined by the formula

$$S = -k \int D \ln (\Gamma D) \, d\Omega, \qquad (1)$$

where k is a universal constant (the Boltzmann constant):

$$k = 1,3804 \cdot 10^{-16} \text{ erg deg}^{-1}.$$

It is mainly there to ensure that thermodynamic entropy and statistical entropy are expressed in the same units. The sign in front of the expression plays a similar part with respect to the sign of the entropy. Γ is a positive quantity which is independent of the phase. Its definition is such that the product

$$\Gamma D$$

is dimensionless. Γ therefore has the same dimensions as Ω. Lastly it is such that, when the system can be broken down into two distinct parts, B and C in accordance with the rules in section 1,

$$\Gamma = \Gamma_B \Gamma_C. \qquad (2)$$

The definition we have just given is therefore a little vague as regards Γ. The reason for this is that any good definition of entropy belongs to quantum physics which supplies an unambiguous definition. Without having recourse to this field we cannot provide as clearcut a definition of entropy as would be wished.

We note in any case that the factor Γ does not come into the variations of the entropy. Formula (1) can in fact be written just as well as

$$k^{-1}S = -\int D \ln D \, d\Omega - \int D \ln \Gamma \, d\Omega, \tag{3}$$

or, since the density in phase is normalized,

$$k^{-1}S = -\int D \ln D \, d\Omega - \ln \Gamma, \tag{4}$$

an expression from which we see that $\ln \Gamma$ enters only as an additive constant.

We should note the clearly unusual nature of entropy as a quantity calculated as a function of the density in phase: as a general rule the density in phase only appears linearly in our formulae.

3.3. Fundamental Inequalities for the Entropy

These inequalities are derived from the properties of the function

$$f(x) = -x \ln x \tag{1}$$

whose first derivative is

$$df/dx = -(\ln x + 1) \tag{2}$$

and the second derivative is

$$d^2f/dx^2 = -1/x. \tag{3}$$

We are interested only in the following values of x:

$$0 \leqq x,$$

since the density in phase is never negative. The second derivative is therefore always negative. Figure 3.1 shows the variation of $f(x)$ as a function of x and also that of

$$g(x) = 1 - x.$$

This figure justifies the inequality

$$-x \ln x \leqq 1 - x, \tag{4}$$

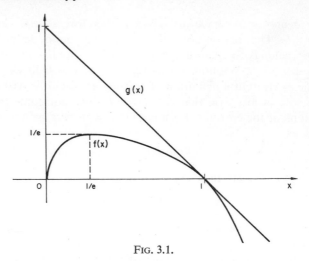

Fig. 3.1.

the equality corresponding only to the case

$$x = 1.$$

Let us apply the inequality (4) to the product aD, D being a density in phase and a a positive coefficient which we shall choose as appropriate. We obtain

$$-aD \ln aD \leqq 1 - aD.$$

We multiply by $d\Omega$ and integrate over a certain finite domain of the extension in phase—and not over the whole extension in phase. We obtain

$$- \int aD \ln aD \, d\Omega \leqq \int d\Omega - a \int D \, d\Omega$$

or

$$- \int D \ln D \, d\Omega \leqq \frac{1}{a} \int d\Omega + \ln a - \int D \, d\Omega.$$

Let us assume that D is zero outside the chosen domain: this system is confined within a certain region of the extension in phase. Then

$$\int D \, d\Omega = 1.$$

Let us make the value of a

$$a = \int d\Omega.$$

62

We obtain
$$- \int D \ln D \, d\Omega \leqq \ln \int d\Omega. \tag{5}$$

The entropy has an upper limit which is reached if
$$aD = 1,$$
i.e. if
$$D = \frac{1}{\int d\Omega}, \tag{6}$$

and only in this case. The entropy is maximum when the density in phase is a constant.

Let us take the inequality (4) again and now replace
$$x \text{ by } \frac{x}{y}, \quad \text{with } y > 0.$$

We obtain
$$- \frac{x}{y} \ln \frac{x}{y} \leqq 1 - \frac{x}{y},$$
or
$$-x \ln x + x \ln y \leqq y - x. \tag{7}$$

Let us apply this inequality to a system A composed of two parts B and C. We have available the reduction relations
$$D_B = \int D \, d\Omega_C,$$
$$D_C = \int D \, d\Omega_B,$$

and three normalization relations
$$\int D \, d\Omega = 1, \quad \int D_B d\Omega_B = 1, \quad \int D_C \, d\Omega_C = 1.$$

We calculate the entropy of the system A:
$$k^{-1}S = - \int D \ln D \, d\Omega - \ln \Gamma_B - \ln \Gamma_C, \tag{8}$$

and the entropy of each of the components:
$$k^{-1}S_B = - \int D_B \ln D_B \, d\Omega_B - \ln \Gamma_B,$$
$$k^{-1}S_C = - \int D_C \ln D_C \, d\Omega_C - \ln \Gamma_C. \tag{9}$$

Putting
$$x = D, \quad y = D_B D_C,$$

the inequality (7) gives
$$-D \ln D + D \ln D_B D_C \leqq D_B D_C - D_A.$$

We integrate over the extension in phase, remembering that

$$d\Omega = d\Omega_B \, d\Omega_C.$$

We obtain

$$- \int D \ln D \, d\Omega + \int D(\ln D_B + \ln D_C) \, d\Omega_B \, d\Omega_C \leqq 0.$$

But, for example,

$$\int D \ln D_B \, d\Omega_B \, d\Omega_C = \int D_B \ln D_B \, d\Omega_B,$$

which means that

$$- \int D \ln D \, d\Omega \leqq - \int D_B \ln D_B \, d\Omega_B - \int D_C \ln D_C \, d\Omega_C. \quad (10)$$

Translated into entropy this relation can be written as

$$S \leqq S_B + S_C, \quad (11)$$

which can be expressed in words by saying that the entropy of a system is less than the sum of the entropies of its components. The equality always obtains if, and only if,

$$D = D_B D_C,$$

i.e. when there is no statistical correlation between the components. An entropy correlation can be defined as

$$S - S_B - S_C$$

and it is always negative.

3.4. Evolution of Statistical Entropy

We shall consider an isolated system. In accordance with the Liouville equation its density in phase changes as follows:

$$\frac{\partial D}{\partial t} + \mathscr{L}D = 0.$$

Any function of the density in phase independent of time also satisfies the Liouville equation. This is the case for the function

$$D \ln D.$$

A result of this is that the quantity S obtained from the preceding expression by integration over all the extension in phase is a constant. The entropy of an isolated system is a constant.

Let us now consider two systems B and C making up a total system A. The system A remains isolated. Its entropy is constant. To put the ideas in a concrete form let B and C each be contained in a box. The phases of the operation are as follows:

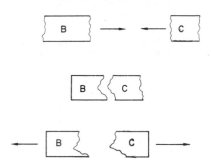

(a) the two systems B and C, which are at first apart, approach each other;

(b) the two systems collide, more or less violently, but retain their individuality;

(c) after a contact of a certain duration the two systems, more or less transformed, rebound and separate.

We formulate an essential hypothesis: before meeting the two systems display no correlation. Therefore,

$$S = S_B + S_C. \tag{1}$$

After meeting a certain correlation

$$S \leqq S_B + S_C \tag{2}$$

will in general be established. But since the total system has remained isolated S is a constant of motion. As a result we have the inequality

$$(S_B + S_C)_{\text{after}} \geqq (S_B + S_C)_{\text{before}}. \tag{3}$$

What measurements can we make? Let us assume that the physicist is able to measure the entropy of B and the entropy of C. The entropy of A is then known before the meeting: we are taking as basic the idea that two systems sent towards each other for the first time have no correlation between them. After the meeting, on the other hand, the entropy of A cannot be measured because we cannot establish the correlations which exist between the two systems B and C; to succeed some connexion at least must be established between B and C, which is contrary to the description of the experiment. All we can do is to measure S_B and S_C separately. The only total entropy that we can measure throughout, i.e.

$$S_B + S_C,$$

cannot therefore have diminished during the meeting.

We know that the entropy of the thermodynamicists has the same properties.

This does not mean that we can identify thermodynamic entropy and statistical entropy because we must now analyse more closely the measurement of S_B or that of S_C. The system B in its turn can be broken down into several elements between which the experimenter will most often have difficulty in establishing the presence or absence of correlation. This means that it is not entirely certain that the measurements about which we speculated above have any physical meaning. To conclude, let us say that this section has simply succeeded in starting to pave the way for comparing the two entropies, a problem which we shall touch on in the last chapter.

3.5. Entropy of a System of Identical Particles

When the system in question is be made up of particles which we shall assume to be identical the entropy can be written explicitly as

$$S = -k \int D_{12\ldots N} \ln(\Gamma_N D_{12\ldots N}) \, d\Omega_{12\ldots N}, \tag{1}$$

the factor Γ_N still having to be determined. In the case when there is doubt in the determination of the occupation it is natural to generalize the above formula by summing over the occupation number:

$$S = -k \sum_N \int D_{12\ldots N} \ln(\Gamma_N D_{12\ldots N}) \, d\Omega_{12\ldots N}. \tag{2}$$

Formula (1) is obviously a special case of this formula (Stratonovich, 1955). When the system is absolutely empty we have simply

$$S = -k D_0 \ln(\Gamma_0 D_0).$$

Since $D_0 = 1$ we make Γ_0 equal to unity so that the entropy is zero.

Justification of formula (2) will appear later. To start with, a study of the perfect fluid will allow us to fix the value of Γ_N.

3.6. Entropy of a Perfect Fluid

Let us calculate the entropy of a perfect fluid. In accordance with section 2.7 we have

$$k^{-1}S = -\sum_0^\infty \int \frac{1}{N!} D_0\mu_1\mu_2 \dots \mu_N$$

$$\times \ln\left(\frac{\Gamma_N}{N!} D_0\mu_1\mu_2 \dots \mu_N\right) d\Omega_{12\dots N}, \tag{1}$$

or, by expanding the logarithm,

$$k^{-1}S = -D_0 \left\{ \ln D_0 \sum_0^\infty \frac{1}{N!} \left(\int \mu_1 \, d\Omega_1\right)^N \right.$$

$$+ \sum_0^\infty \frac{1}{N!} \left(\int \mu_1 \, d\Omega_1\right)^N \ln \frac{\Gamma_N}{N!}$$

$$\left. + \int \mu_1 \ln \mu_1 \, d\Omega_1 \sum_1^\infty \frac{1}{(N-1)!} \left(\int \mu_1 \, d\Omega_1\right)^{N-1} \right\}. \tag{2}$$

But

$$\sum_0^\infty \frac{1}{N!} \left(\int \mu_1 \, d\Omega_1\right)^N = \sum_1^\infty \frac{1}{(N-1)!} \left(\int \mu_1 \, d\Omega_1\right)^{N-1}$$

$$= \exp\left(\int \mu_1 \, d\Omega_1\right) = D_0^{-1}.$$

Therefore,

$$k^{-1}S = \int \mu_1(1 - \ln \mu_1) \, d\Omega_1$$

$$- \exp\left(-\int \mu_1 \, d\Omega_1\right) \sum_0^\infty \frac{1}{N!} \left(\int \mu_1 \, d\Omega_1\right)^N \ln \frac{\Gamma_N}{N!}. \tag{3}$$

This result allows us to determine Γ_N. It is convenient in fact that in this correlationless fluid the entropy should be the sum of the entropies of each element of volume. We can obtain this result only by putting

$$\Gamma_N = \Gamma^N N!, \tag{4}$$

Γ remaining undetermined. This result is compatible with the value that we have taken for Γ_0. We now have

$$k^{-1}S = \int \mu_1(1 - \ln \Gamma\mu_1) \, d\Omega_1. \tag{5}$$

We have already introduced in section 2.5 a factor Γ which played a dimensional part like the one we have just introduced. It is natural to identify them with one another. But whilst the choice that we made above,

$$\Gamma = h^3, \tag{6}$$

was arbitrary, it is now imposed by quantum theory. Classical theory which we have followed up to now is incapable of determining it. But the classical theory is an approximation of quantum theory, an approximation which is valid when Planck's constant is negligible: in certain cases, of course, Planck's constant cannot be completely eliminated. We assume in this volume the quantum result (6) without any justification. The smallness of Γ, from the point of view of the classical theory, means that the inequality

$$\Gamma \mu_1 \ll 1$$

is easy to satisfy in the useful field of application of this theory. As a consequence the second term of the entropy (5) is always positive, like the first.

3.7. Fundamental Inequality for Coupled Systems

Let us divide the space occupied by a system A made up of identical particles into two parts B and C. The problem is to find out whether the fundamental inequality is still valid in this case.

We assume in general that the occupation of A is poorly determined and likewise that of B and of C. The study will cover all the special cases that can be thought of:

occupation of A fixed,

occupation of B fixed,

occupations of A, of B and of C simultaneously fixed (this case does not differ from that which was dealt with in section 2.3), and also the case of systems which are not coupled— but which, for example, could be coupled or could have been coupled at another point in time.

We shall use

$$D_0, D_1, D_{12}, ..., D_{12...N}, ...$$

to denote the densities of occupation of A, and

$$\pi_0 = D_0, \pi_1, ..., \pi_N, ...$$

to denote the corresponding probabilities of occupation. The same quantities relating to B or C will be characterized by the suffix B or C. Lastly the probability of simultaneously having M particles in B and P particles in C will be denoted by $\pi_{M,P}$. As regards these probabilities we have available the following relations:

$$\sum_N \pi_N = 1,$$

$$\sum_N \pi_{BN} = 1,$$

$$\sum_N \pi_{CN} = 1,$$

$$\pi_{BN} = \sum_P \pi_{N,P}, \quad \pi_{CN} = \sum_M \pi_{M,N},$$

$$\pi_N = \sum_{\substack{M=0 \\ P=N-M}}^{N} \pi_{M,P}. \tag{1}$$

In addition, we can calculate the $\pi_N, \pi_{BN}, \pi_{CN}$ as functions of the corresponding densities of occupation. We now have to express $D_{B12...M}$, for example, as a function of $D_{12...N}$.

Let us first discuss the case when the occupation of A is well known and equal to N. Of course, the quantity being studied is zero as soon as we have

$$M > N.$$

Otherwise we have

$$D_{B12...M} = \frac{N!}{M!(N-M)!} \int_C D_{12...N} \, d\Omega_{M+1,\,...,\,N}. \tag{2}$$

Here the first M particles are part of B, whilst the remaining particles are part of C. As far as the factorials are concerned, they are necessary to express that it does not matter whether the first M particles are the ones that are in B: any other set of M particles makes physically the same contribution.

When the occupation of A is poorly known formula (2) can be generalized easily. It is sufficient to add the contributions to D_B obtained from all the occupations of A:

$$D_{B12...M} = \sum_N \frac{N!}{M!(N-M)!} \int_C D_{12...N} \, d\Omega_{M+1,\,...,\,N}. \tag{3}$$

Statistical Entropy

A similar procedure can be applied to calculating the integral

$$I_N = - \int D_{12\ldots N} \ln D_{12\ldots N} \, d\Omega_{12\ldots N} \tag{4}$$

which can be made more precise when we know the number of particles which are in B. We have

$$I_N = - \sum_M \frac{N!}{M!(N-M)!} \int_B \int_C D_{12\ldots N} \ln D_{12\ldots N}$$

$$\times \, d\Omega_{12\ldots M} \, d\Omega_{M+1,\ldots,N}. \tag{5}$$

After this operation the entropy of A can be put in the following form:

$$k^{-1}S_A = - \sum_M \sum_P \int_B \int_C \frac{(M+P)!}{M!P!} D_{1,2,\ldots,M+P}$$

$$\times \ln \frac{(M+P)!}{M!P!} D_{1,2,\ldots,M+P} \, d\Omega_{12,\ldots,M} \, d\Omega_{M+1,\ldots,M+P}$$

$$+ \sum_M \sum_P \frac{(M+P)!}{M!P!} \ln \frac{(M+P)!}{M!P!} \int_B \int_C D_{12,\ldots,M+P}$$

$$\times \, d\Omega_{12,\ldots,M} \, d\Omega_{M+1,\ldots,M+P} - \sum_N \pi_N \ln \Gamma_N. \tag{6}$$

The expression

$$\frac{(M+P)!}{M!P!} \int_B \int_C D_{12,\ldots,M+P} \, d\Omega_{12\ldots M} \, d\Omega_{M+1,\ldots,M+P} \tag{7}$$

is nothing other than $\pi_{M,P}$. We therefore rewrite the entropy:

$$k^{-1}S_A = - \sum_M \sum_P \int_B \int_C \frac{(M+P)!}{M!P!} D_{1,2,\ldots,M+P}$$

$$\times \ln D_{1,2,\ldots,M+P} \, d\Omega_{12\ldots M} \, d\Omega_{M+1,\ldots,M+P}$$

$$+ \sum_M \sum_P \pi_{M,P} \ln \frac{(M+P)!}{M!P!} - \sum_N \pi_N \ln \Gamma_N. \tag{8}$$

We can now make use of the inequality

$$-x \ln x \leq -x \ln y + y - x$$

70

by putting

$$x = \frac{(M-P)!}{M!P!} D_{1,2,\ldots,M+P},$$

$$y = D_{B^{12}\ldots M} D_{C^{M+1},\ldots,M+P}. \qquad (9)$$

We obtain, bearing equation (2) in mind,

$$k^{-1}S_A \leqq -\sum_M \int_B D_{B^{12}\ldots M} \ln D_{B^{12}\ldots M} \, d\Omega_{12\ldots M}$$

$$-\sum_P \int_C D_{C^{12}\ldots P} \ln D_{C^{12}\ldots P} \, d\Omega_{12\ldots P}$$

$$+\sum_M \sum_P (\pi_{B^M} \pi_{C^P} - \pi_{M,P})$$

$$+\sum_M \sum_P \pi_{M,P} \ln \frac{(M+P)!}{M!P!} - \sum_N \pi_N \ln \Gamma_N. \qquad (10)$$

The last three terms can be simplified because

$$\sum_M \sum_P \pi_{B^M} \pi_{C^P} = \sum_M \pi_{B^M} \sum_P \pi_{C^P} = 1, \qquad \sum_M \sum_P \pi_{M,P} = 1,$$

$$\sum_M \sum_P \pi_{M,P} \ln \frac{(M+P)!}{M!P!} = \sum_N \pi_N \ln N! - \sum_M \pi_M \ln M! - \sum_P \pi_P \ln P!,$$

$$\sum_N \pi_N \ln \Gamma_N = \sum_N \pi_N \ln N! + \sum_N \pi_N \ln \Gamma^N,$$

$$\sum_N \pi_N \ln \Gamma_N = \sum_M \pi_M \ln \Gamma_M + \sum_P \pi_P \ln \Gamma^P.$$

Finally we obtain the generalization of the fundamental inequality:

$$S_A \leqq S_B + S_C. \qquad (11)$$

It justifies the choice of the definition (2) of section 3.5.

3.8. Entropy Expressed as a Function of Reduced Densities [†]

Let us return to the h_1, h_{12}, h_{123} which we used in section 2.5 for analysing the correlations. The entropy (2) of section 3.5 can

[†] The expressions obtained in this section are due to Nettleton and Green (1958).

then be rewritten as follows:

$$k^{-1}S = -\left(D_0 + \int D_1 \, d\Omega_1 + \int D_{12} \, d\Omega_{12} + \cdots\right) \ln D_0$$
$$- \int D_1 h_1 \, d\Omega_1 + \int D_{12}(h_1 + h_2 + h_{12}) \, d\Omega_{12}$$
$$- \int D_{123}(h_1 + h_2 + h_3 + h_{12} + h_{13} + h_{23} + h_{123}) \, d\Omega_{123}$$
$$+ \cdots \tag{1}$$

or, bearing in mind the symmetry and the normalization,

$$k^{-1}S = -\ln D_0$$
$$- \int h_1 \left(D_1 + 2 \int D_{12} \, d\Omega_2 + 3 \int D_{123} \, d\Omega_{23} + \cdots\right) d\Omega_1$$
$$- \int h_{12} \left(D_{12} + 3 \int D_{123} \, d\Omega_3 + 4 \int D_{1234} \, d\Omega_{34} + \cdots\right) d\Omega_{12}$$
$$+ \cdots . \tag{2}$$

We arrive at the following expression:

$$k^{-1}S = -\ln D_0 - \int \mu_1 h_1 \, d\Omega_1 - \frac{1}{2!} \int \mu_{12} h_{12} \, d\Omega_{12}$$
$$- \frac{1}{3!} \int \mu_{123} h_{123} \, d\Omega_{123} - \cdots, \tag{3}$$

where the μ derive from the brackets. It is elegant but mixes different kinds of quantities. We must go further and eliminate every trace of the densities of occupation so as not to allow anything but the reduced densities to appear. This undertaking is going to lead us to calculations which are a little long but are important. We first must go back to formulae (12), (13) and (14) of section 2.6 which express the densities of occupation in terms of the reduced densities and of D_0, a quantity which we shall retain for the moment. For example, we can rewrite equation (12) by using the parameters a defined in section 2.8. We have:

$$D_1 = D_0 \mu_1 \left\{ 1 - \int \mu_2 (a_{12} - 1) \, d\Omega_2 \right.$$
$$+ \tfrac{1}{2} \int \mu_2 \mu_3 (a_{12} a_{13} a_{23} a_{123} - a_{12} - a_{13} - a_{23} + 2) \, d\Omega_{23}$$
$$- \tfrac{1}{6} \int \mu_2 \mu_3 \mu_4 (a_{12} a_{13} a_{14} a_{23} a_{24} a_{34} a_{123} a_{124} a_{134} a_{234} a_{1234}$$
$$- a_{12} a_{13} a_{23} a_{123} - a_{12} a_{14} a_{24} a_{124} - a_{13} a_{14} a_{34} a_{134}$$
$$- a_{23} a_{24} a_{34} a_{234} - a_{12} a_{34} - a_{13} a_{24} - a_{14} a_{23} + 2a_{12}$$
$$\left. + 2a_{13} + 2a_{14} + 2a_{23} + 2a_{24} + 2a_{34} - 6) \, d\Omega_{234} + \cdots \right\}.$$

Entropy Expressed as a Function of Reduced Densities

To obtain h we must calculate the logarithm of the large bracket. We can again use the cumulant method here. The suffix 1 plays no part in this calculation. We have

$$\frac{D_1}{D_0} = \Gamma^{-1} \exp h_1 = \mu_1 \exp\left\{- \int \mu_2(a_{12} - 1)\, d\Omega_2\right.$$

$$+ \frac{1}{2} \int \mu_2\mu_3(a_{12}a_{13}a_{23}a_{123} - a_{12}a_{13} - a_{23} + 1)\, d\Omega_{23}$$

$$- \frac{1}{6} \int \mu_2\mu_3\mu_4(a_{12}a_{13}a_{14}a_{23}a_{24}a_{34}a_{123}a_{124}a_{134}a_{234}a_{1234}$$

$$- a_{12}a_{13}a_{23}a_{123}a_{14} - a_{12}a_{14}a_{24}a_{124}a_{13}$$

$$- a_{13}a_{14}a_{34}a_{134}a_{12} - a_{23}a_{24}a_{34}a_{234} + 2a_{12}a_{13}a_{14}$$

$$\left. + a_{23} + a_{24} + a_{34} - 2)\, d\Omega_{234} + \cdots\right\}; \tag{4}$$

and likewise

$$\frac{D_{12}}{D_0} = \frac{1}{2}\Gamma^{-2} \exp\left(h_1 + h_2 + h_{12}\right)$$

$$= \frac{1}{2}\mu_{12} \exp\left\{- \int \mu_3(a_{13}a_{23}a_{123} - 1)\, d\Omega_3\right.$$

$$+ \frac{1}{2} \int \mu_3\mu_4(a_{13}a_{14}a_{23}a_{24}a_{34}a_{123}a_{124}a_{134}a_{234}a_{1234}$$

$$\left. - a_{13}a_{23}a_{123}a_{14}a_{24}a_{124} - a_{34} + 1)\, d\Omega_{34} - \cdots\right\}, \tag{5}$$

$$\frac{D_{123}}{D_0} = \frac{1}{6}\Gamma^{-3} \exp\left(h_1 + h_2 + h_3 + h_{12} + h_{13} + h_{23} + h_{123}\right)$$

$$= \frac{1}{6}\mu_{123} \exp\left\{- \int \mu_4(a_{14}a_{24}a_{34}a_{124}a_{134}a_{234}a_{1234} - 1)\right.$$

$$\left. \times d\Omega_4 + \cdots\right\},$$

$$\frac{D_{1234}}{D_0} = \frac{1}{24}\Gamma^{-4} \exp\left(h_1 + \cdots + h_{1234}\right)$$

$$= \frac{1}{24}\mu_{1234} \exp\left\{- \int \cdots d\Omega_5 + \cdots\right\}.$$

These results prove that the densities of occupation remain positive whatever the reduced densities, provided they are reduced equally (and that the expansions have a meaning!). In each case we have limited the expansions to terms with four suffices. From this we obtain

$$h_1 = \ln \Gamma \mu_1 - \int \mu_2 (a_{12} - 1) \, d\Omega_2$$

$$+ \tfrac{1}{2} \int \mu_2 \mu_3 (a_{12} a_{13} a_{23} a_{123} - a_{12} a_{13} - a_{23} + 1) \, d\Omega_{23}$$

$$- \tfrac{1}{6} \int \mu_2 \mu_3 \mu_4 (a_{12} a_{13} a_{14} a_{23} a_{24} a_{34} a_{123} a_{124} a_{134} a_{234} a_{1234}$$

$$- a_{12} a_{13} a_{23} a_{123} a_{14} - a_{12} a_{14} a_{24} a_{124} a_{13} - a_{13} a_{14} a_{34} a_{134} a_{12}$$

$$- a_{23} a_{24} a_{34} a_{234} + 2 a_{12} a_{13} a_{14} + a_{23} + a_{24} + a_{34} - 2)$$

$$\times \, d\Omega_{234} + \cdots; \tag{6}$$

$$h_{12} = \ln a_{12} - \int \mu_3 (a_{13} a_{23} a_{123} - a_{13} - a_{23} + 1) \, d\Omega_3$$

$$+ \tfrac{1}{2} \int \mu_3 \mu_4 (a_{13} a_{14} a_{23} a_{24} a_{34} a_{123} a_{124} a_{134} a_{234} a_{1234}$$

$$- a_{13} a_{23} a_{123} a_{14} a_{24} a_{124} - a_{13} a_{14} a_{34} a_{134} - a_{23} a_{24} a_{34} a_{234}$$

$$+ a_{13} a_{14} + a_{23} a_{24} + a_{34} - 1) \, d\Omega_{34} - \cdots;$$

$$h_{123} = \ln a_{123} - \int \mu_4 (a_{14} a_{24} a_{34} a_{124} a_{134} a_{234} a_{1234}$$

$$- a_{14} a_{24} a_{124} - a_{14} a_{34} a_{134}$$

$$- a_{24} a_{34} a_{234} + a_{14} + a_{24} + a_{34} - 1) \, d\Omega_4 + \cdots$$

$$h_{1234} = \ln a_{1234} - \cdots.$$

We substitute these expressions in the expansion (3) of the entropy. At the same time we replace $\ln D_0$ by its expression (9) of section 2.6 which we transcribe as

$$\ln D_0 = - \int \mu_1 \, d\Omega_1 + \frac{1}{2} \int \mu_1 \mu_2 (a_{12} - 1) \, d\Omega_{12}$$

$$- \frac{1}{6} \int \mu_1 \mu_2 \mu_3 (a_{12} a_{13} a_{23} a_{123} - a_{12} - a_{13} - a_{23} + 2)$$

$$\times \, d\Omega_{123} + \frac{1}{24} \int \mu_1 \mu_2 \mu_3 \mu_4 (a_{12} a_{13} a_{14} a_{24} a_{23} a_{34} a_{123}$$

$$\times \, a_{124} a_{134} a_{234} a_{1234} - a_{12} a_{13} a_{23} a_{123} - a_{12} a_{14} a_{24} a_{124}$$

$$- a_{13} a_{14} a_{34} a_{134} - a_{23} a_{24} a_{34} a_{234} - a_{12} a_{34} - a_{13} a_{24}$$

$$- a_{14} a_{23} + 2 a_{12} + 2 a_{13} + 2 a_{14} + 2 a_{23} + 2 a_{24}$$

$$+ 2 a_{34} - 6) \, d\Omega_{1234} - \cdots. \tag{7}$$

After ordering it according to the number of suffices we obtain

$$k^{-1}S = \int \mu_1(1 - \ln \Gamma\mu_1)\, d\Omega_1$$

$$- \frac{1}{2} \int [\mu_{12} \ln a_{12} - \mu_1\mu_2(a_{12} - 1)]\, d\Omega_{12}$$

$$- \frac{1}{6} \int [\mu_{123} \ln a_{123} - \mu_1\mu_2\mu_3(a_{12}a_{13}a_{23}a_{123} - a_{12}a_{13}$$

$$- a_{12}a_{23} - a_{13}a_{23} + a_{12} + a_{13} + a_{23} - 1)]\, d\Omega_{123}$$

$$- \frac{1}{24} \int [\mu_{1234} \ln a_{1234} - \mu_1\mu_2\mu_3\mu_4(a_{12}a_{13}a_{14}a_{23}a_{24}a_{34}$$

$$\times\ a_{123}a_{124}a_{134}a_{234}a_{1234} - a_{12}a_{13}a_{23}a_{123}a_{14}a_{24}a_{124}$$

$$- a_{13}a_{12}a_{23}a_{123}a_{14}a_{34}a_{134}$$

$$- a_{14}a_{12}a_{24}a_{124}a_{13}a_{34}a_{134}$$

$$- a_{23}a_{12}a_{13}a_{123}a_{24}a_{34}a_{234}$$

$$- a_{24}a_{12}a_{14}a_{124}a_{23}a_{34}a_{234}$$

$$- a_{34}a_{13}a_{14}a_{134}a_{23}a_{24}a_{234} + a_{12}a_{13}a_{23}a_{123}a_{14}$$

$$+ a_{12}a_{14}a_{24}a_{124}a_{13} + a_{13}a_{14}a_{34}a_{134}a_{12}$$

$$+ a_{12}a_{13}a_{23}a_{123}a_{24} + a_{12}a_{14}a_{24}a_{124}a_{23}$$

$$+ a_{23}a_{24}a_{34}a_{234}a_{12} + a_{12}a_{13}a_{23}a_{123}a_{34}$$

$$+ a_{13}a_{14}a_{34}a_{134}a_{23} + a_{23}a_{24}a_{34}a_{234}a_{13}$$

$$+ a_{12}a_{14}a_{24}a_{124}a_{34} + a_{13}a_{14}a_{34}a_{134}a_{24}$$

$$+ a_{23}a_{24}a_{34}a_{234}a_{14} - a_{12}a_{13}a_{23}a_{123}$$

$$- a_{12}a_{14}a_{24}a_{124} - a_{13}a_{14}a_{34}a_{134}$$

$$- a_{23}a_{24}a_{34}a_{234} - a_{12}a_{13}a_{14} - a_{12}a_{23}a_{24}$$

$$- a_{13}a_{23}a_{34} - a_{14}a_{24}a_{34} - a_{12}a_{34} - a_{13}a_{24}$$

$$- a_{14}a_{23} + a_{12} + a_{13} + a_{14} + a_{23} + a_{24} + a_{34} - 2)]$$

$$\times d\Omega_{1234} - \cdots. \tag{8}$$

To obtain this perfectly symmetrical result it has been necessary to juggle with the choice of suffices.

When there is no need for the symmetry to be explicit we keep only one term of each type with an appropriate coefficient.

We then get

$$k^{-1}S = \int \mu_1(1 - \ln \Gamma\mu_1) \, \mathrm{d}\Omega_1$$

$$+ \frac{1}{2} \int \left[-\mu_{12} \ln a_{12} + \mu_1\mu_2(a_{12} - 1) \right] \mathrm{d}\Omega_{12}$$

$$+ \frac{1}{6} \int \left[-\mu_{123} \ln a_{123} + \mu_1\mu_2\mu_3(a_{12}a_{13}a_{23}a_{123} \right.$$

$$\left. - 3a_{12}a_{13} + 3a_{12} - 1) \right] \mathrm{d}\Omega_{123}$$

$$+ \frac{1}{24} \int \left[-\mu_{1234} \ln a_{1234} + \mu_1\mu_2\mu_3\mu_4(a_{12}a_{13}a_{14}a_{23}a_{24} \right.$$

$$\times a_{34}a_{123}a_{124}a_{134}a_{234}a_{1234} - 6a_{12}a_{13}a_{23}a_{123}a_{14}a_{24}a_{124}$$

$$+ 12a_{12}a_{14}a_{24}a_{124}a_{13} - 4a_{12}a_{13}a_{23}a_{123} - 4a_{12}a_{13}a_{14}$$

$$\left. - 3a_{12}a_{34} + 6a_{12} - 2) \right] \mathrm{d}\Omega_{1234} - \cdots \qquad (9)$$

By introducing the ε we finally bring out the short range of the integrals in the case when molecular disorder obtains:

$$k^{-1}S = \int \mu_1(1 - \ln \Gamma\mu_1) \, \mathrm{d}\Omega_1$$

$$+ \frac{1}{2} \int \left[-\mu_{12} \ln a_{12} + \mu_1\mu_2\varepsilon_{12} \right] \mathrm{d}\Omega_{12}$$

$$+ \frac{1}{6} \int \left[-\mu_{123} \ln a_{123} + \mu_1\mu_2\mu_3(a_{12}a_{13}a_{23}\varepsilon_{123} \right.$$

$$\left. + \varepsilon_{12}\varepsilon_{13}\varepsilon_{23}) \right] \mathrm{d}\Omega_{123}$$

$$+ \frac{1}{24} \int \left[-\mu_{1234} \ln a_{1234} + \mu_1\mu_2\mu_3\mu_4(a_{12}a_{13}a_{14}a_{23}a_{24} \right.$$

$$\times a_{34}a_{123}a_{124}a_{134}a_{234}\varepsilon_{1234} + a_{12}a_{13}a_{14}a_{23}a_{24}a_{34}\varepsilon_{123}\varepsilon_{124}$$

$$\times \varepsilon_{134}\varepsilon_{234} + 4a_{12}a_{13}a_{14}a_{23}a_{24}a_{34}\varepsilon_{123}\varepsilon_{124}\varepsilon_{134}$$

$$+ 6a_{12}a_{13}a_{14}a_{23}a_{24}\varepsilon_{34}\varepsilon_{123}\varepsilon_{124} + 4\varepsilon_{14}\varepsilon_{24}\varepsilon_{34}a_{12}a_{13}a_{23}\varepsilon_{123}$$

$$\left. + \varepsilon_{12}\varepsilon_{13}\varepsilon_{14}\varepsilon_{23}\varepsilon_{24}\varepsilon_{34} - 3\varepsilon_{12}\varepsilon_{14}\varepsilon_{23}\varepsilon_{34}) \right] \mathrm{d}\Omega_{1234} - \cdots. \qquad (10)$$

Once again the terms are in the order of the number of suffices they have. In the case of a perfect fluid the general expression (10) is reduced to its first term. It is the same as the result previously obtained.

If the perfect fluid hypothesis reduces the entropy to its first term, the same is not true of Kirkwood's approximation which involves in principle the appearance all the higher-order terms.

The a's (and therefore the ε's) being given, the entropy becomes a functional of the μ_i which is expressed in terms of increasing powers. Under these conditions let us make the μ_i's approach zero, which reduces the occupation correspondingly. It is obvious that an approximate expression can be obtained for the entropy by retaining only the first terms. The approximation boils down basically to assuming that the fluid is perfect. If we move on to the second order we must add to the entropy calculated above the quantity

$$\tfrac{1}{2} \int (-\mu_{12} \ln a_{12} + \mu_1\mu_2\varepsilon_{12}) \, d\Omega_{12}. \tag{11}$$

This quantity is necessarily negative because this is the case for the integrand. Let us arbitrarily divide the total system into two parts B and C. A result of the preceding remark is that we have

$$S_A \leqq S_B + S_C, \tag{12}$$

the equality holding in the case when ε_{12} is zero as soon as the two particles are not both in either B or C.

The generalization of this result to higher orders looks difficult. We must remember that here we have established the general validity of the inequality (12) in section 7. We can at least use formula (10) to verify that the entropy is additive when there is no correlation between B and C.

3.9. Applications

On the one hand, the entropy, in its form (1) of section 2 or (2) of section 5, will serve us to introduce the laws of thermodynamic equilibrium on the basis of a unique principle.

On the other hand, in its form (10) of section 8, it will allow us to touch on questions relating to Carnot's principle.

CHAPTER 4

Canonical Laws of Thermodynamic Equilibrium

4.1. Introduction

Let us consider a macroscopic physical system, i.e. one with a very large number of degrees of freedom. It is contained in a box at rest, or, to use more theoretical language, is confined in a fixed force field. It is isolated: it does not exchange energy or particles with its surroundings. We know that any natural system thus left to itself ends up by being stabilized in a state of equilibrium which is called thermodynamic equilibrium.

In this chapter we shall refrain completely from trying to find out how this equilibrium is set up little by little. We shall only try to describe it.

The density in phase of a system at rest satisfies the equation

$$\mathscr{L}D = 0. \tag{1}$$

Except for special cases, which we shall neglect, the generally suitable solution of this equation is a function of the Hamiltonian*

$$D = D(H). \tag{2}$$

But thermodynamic equilibrium is not defined only by the condition (1) and it is more restrictive than indicated by the formula (2).

The principle which allows us to formulate thermodynamic equilibrium is as follows: bearing in mind the restrictions imposed, the entropy of a system in thermodynamic equilibrium is maximal.

The following examples will show which restraints are meant. No more is ascribed to the statement above than it contains. It does not state in any way how the equilibrium can be achieved.

---

* For a discussion of this point see ter Haar (1966).

4.2. Microcanonical Equilibrium

We specify that the representative point of the system in phase space is subject to the condition

$$E < H < E + \Delta E,$$

the energy range ΔE being very small when compared with E. We assume that the region of extension in phase thus defined is finite. The restraints having thus been fixed the calculation of the density in phase which ensures the maximum of the entropy has already been made in connexion with the general study of the latter. The solution is as follows:

$$D(H) = \frac{1}{\int_{E}^{E+\Delta E} d\Omega} \quad \text{for} \quad E < H < E + \Delta E;$$

$D(H) = 0$ when E lies outside the range $E, E + \Delta E$.

The value of the entropy is

$$k^{-1}S = \ln \frac{\int d\Omega}{\Gamma}.$$

4.3. Canonical Equilibrium

Instead of limiting the possible values of the energy to a narrow range we prefer to fix simply the average energy U. The following relations are available to us:

$$k^{-1}S = -\int D \ln D \, d\Omega - \ln \Gamma,$$

$$1 = \int D \, d\Omega,$$

$$U = \int DH \, d\Omega.$$

We shall show that the density in phase D given by the equation

$$D = \frac{1}{\xi} e^{-\beta H} \tag{1}$$

gives the answer to our problem. Here, β is a parameter which is necessarily positive in order that the proposed density in phase can be normalized. Moreover, it is necessary that the Hamiltonian function is bounded from below.

The parameter ξ is thus determined by the normalization condition

$$\xi = \int e^{-\beta H} \, d\Omega, \tag{2}$$

and it is thus a function of β and, of course, of the characteristic parameters of the system.

The parameter β itself is determined from the condition on the energy,

$$U = \int HD \, d\Omega. \tag{3}$$

This condition is equivalent to the following one:

$$\frac{d \ln \xi}{d\beta} = U. \tag{4}$$

Let us assume, to fix the ideas, that the lower bound of H is equal to zero. The derivative on the left-hand side of (4) is always positive. It vanishes when β is infinite, and is infinite when β vanishes. The problem has always one and only one solution.

The statistical entropy corresponding to the situation represented by the density in phase D of (1) is given by the following equation:

$$S/k = \ln (\xi/\Gamma) + \beta U. \tag{5}$$

We must thus prove that this entropy is a maximum under the given constraint.

Let us therefore consider an arbitrary density in phase D' which is not even necessarily independent of the time and let us express the corresponding entropy as follows:

$$S'/k = - \int D' \ln (\Gamma D') \, d\Omega.$$

We have always

$$-D' \ln D' + D' \ln D \leqq D - D',$$

and hence*

$$S'/k \leqq - \int D' \ln (\Gamma D) \, d\Omega, \tag{6}$$

or, using the definition (1),

$$S'/k \leqq \ln (\Gamma/\xi) + \beta \int D'H \, d\Omega.$$

If we now impose upon the density in phase D' the same condition (3) of a given value of the average energy, we have

$$S' \leqq S. \tag{7}$$

The distribution law (1) realizes thus the entropy maximum. It is the "Boltzmann–Gibbs distribution law" or the "canonical distribution law".

The physical significance of β will become clear to us from the following reasoning. Let us consider a system made up out of two parts B and C separated by a partition through which energy can

---

* This inequality is also in exactly the same form valid in quantum theory.

pass. The Hamiltonian of the system is

$$H = H_B + H_C + H_I.$$

We shall assume that H_I can be neglected in equilibrium. In any case this weak coupling ensures that the two systems B and C are thermalized together. We fix the average total energy. The density in phase is therefore

$$D = \frac{1}{\xi} e^{-\beta(H_B + H_C)}. \tag{8}$$

There are no correlations between the two systems. Each of them corresponds to a canonical distribution

$$D_B = \frac{1}{\xi_B} e^{-\beta H_B}, \tag{9}$$

$$D_C = \frac{1}{\xi_C} e^{-\beta H_C},$$

the parameter β being the same for B and for C. Experience tells us that two such systems are at the same temperature. We are therefore led to establish a general kind of relation between β and the temperature:

$$\beta = f(T).$$

This correspondence should be unambiguous. Since high temperatures favour high energies, β and T should vary in opposite directions. We therefore put, a more complete justification to be supplied *a posteriori*,

$$\beta = \frac{1}{kT}. \tag{10}$$

We should point out that the temperature in question here is the absolute temperature of the thermodynamicists. As for k, it is the Boltzmann constant which we have already met when dealing with the entropy. As is appropriate, kT has the dimensions of an energy.

Relation (7) can now be put in the advantageous form

$$\ln \frac{\Gamma}{\xi} = \frac{U - TS}{kT}. \tag{11}$$

4.4. Grand Canonical Equilibrium

We return now to the idea that in general we do not know exactly how many particles the system contains.

We shall restrict ourselves here, however, to the case where there is only a single kind of particle.

When there are N particles present $d\Omega_N$ is the element of extension in phase, D_N the corresponding density in phase and of occupation. The particles are then numbered from 1 to N.

The probability that there are N particles present is denoted by π_N. The π_N should satisfy the normalization conditions

$$\sum_N \pi_N = 1.$$

The expression for the entropy is

$$k^{-1}S = -\sum_N \int D_N \ln(\Gamma_N D_N) \, d\Omega_N. \tag{1}$$

The thermodynamic equilibrium that we obtain by saying that this entropy is maximal takes into account two restrictions:

The value of the average energy is given

$$U = \sum_N \int H_N D_N \, d\Omega_N,$$

and also the average number of particles

$$\langle N \rangle = \sum_N N \int D_N \, d\Omega_N = \sum_N \pi_N N.$$

This equilibrium is called the grand canonical equilibrium.

The method of the preceding section shows that the entropy is maximum when the D_N satisfy the following relation:

$$\ln(\Gamma_N D_N) = -\ln \Xi + \alpha N - \beta H_N, \tag{2}$$

where the meaning of the arbitrary coefficients $\ln \Xi$, α and β should become clear later; α and Ξ are dimensionless numbers.

We have

$$D_N = \frac{1}{\Xi} \frac{e^{\alpha N - \beta H_N}}{\Gamma_N} \tag{3}$$

and, by using the parameter ξ introduced into the theory of canonical equilibrium,

$$\pi_N = \frac{1}{\Xi} \frac{e^{\alpha N} \xi_N}{\Gamma_N}. \tag{4}$$

Ξ is obtained from the normalization condition

$$\Xi = \sum_N \frac{e^{\alpha N} \xi_N}{\Gamma_N}. \tag{5}$$

As for α and β, they are determined by the restrictions. Following the same line of argument as before β should be positive. Formula (1) allows us to calculate the entropy itself. We shall put the relation obtained in the following form:

$$- \ln \varXi = \frac{U - TS - kT\alpha \langle N \rangle}{kT}.$$

(6)

The quantity $\varXi$ is called the grand partition function.

We note that $\varXi$ has a very simple meaning:

$$\varXi = D_0^{-1}.$$

The quantity $\omega = kT\alpha$ is called the chemical potential and β has been identified here also as $1/kT$.

In the study of large systems, systems which have a very high number of degrees of freedom, the different distributions that we have just examined are equivalent.

The simplest way of understanding it without spending a lot of time on this follows from studying the fluctuations.

In a large canonical system the energy fluctuations are in general very small, so much so that to all intents and purposes the energy might be fixed.

In a large grand canonical system the energy and occupation fluctuations are likewise in general very small, so much so that to all intents and purposes the energy and occupation are fixed.

In practice we prefer to work with the most "liberal" distributions, i.e. the canonical distribution when the exact nature of the system is not stated and the grand canonical distribution when we are dealing with systems made up of a large number of identical particles, either of one kind only or of several kinds.

More closely reasoned arguments would make it possible to establish closer connexions between the three kinds of distribution. Despite their interest we shall not dwell upon them here because they are not applicable in the case of long-range forces and Coulomb forces in particular. We shall remember in any case that it is often convenient to incorporate the system under study in a larger system which will itself be in thermodynamic equilibrium.

4.5. Neutral Equilibria

In general our theories of thermodynamic equilibrium will prove satisfactory. There is, however, a case where experiment and theory do not agree and which contains strange features.

Let us imagine the following experiment: a vast, flat-bottomed cylindrical container has vertical walls. It is made of glass, for example, and contains mercury. The glass is homogeneous and

83

perfectly clean. The force of gravity exerted in this container is on the laboratory scale. The temperature is well defined. If the amount of mercury introduced into the container is small the mercury vaporizes completely. If there is more of it drops of mercury form which rest on the bottom but do not spread out because the mercury does not wet the glass. The situation described will evolve little by little to the advantage of the largest drop which will end up by collecting together all the liquid mercury. It will flatten out to a greater or lesser extent because of gravity and will take up a circular shape. If the bottom of the container is much larger than the drop the position of its centre will remain unpredictable in practice. It is possible that as a result of slow fluctuations it will be displaced little by little: this phenomenon will not be open to continuous observation.

The theory of thermodynamic equilibrium does not distinguish between slow fluctuations and rapid fluctuations. In theory all the possible positions of the drop are equally probable—we shall ignore the cases when it would touch the wall. If we ask the theory to evaluate the density of the mercury in the bottom of the container it will give us the average density which could be calculated from experiment by putting the drop in all the possible positions; in practice this density will depend only on the altitude, and will be mixed up with the density of the vapour. Nothing will give us the density of the liquid or the shape of the drop.

There is, fortunately, a way of avoiding the effect of the rigorousness of the theory, which transforms a drop that can be properly observed into a kind of phantom: this is to imagine a small irregularity in the surface of the glass or in the force field which destroys the symmetry of the set-up and which eliminates the neutral nature of the theoretical equilibrium without in any way altering the aspects of the mercury's condensation.

Remarks of the same kind apply to the theory of ferromagnetism: in the absence of an applied field magnetization has no more reason to occur in one direction than in the opposite direction. When the theory is manipulated with care it will produce a magnetization which will be the average of two opposed attractions which are equally probable, i.e. a zero magnetization. This is not what is observed experimentally. To avoid this contradiction the theory is developed to allow for an applied magnetic field: one passes to the limit of a zero magnetic field only with care.

CHAPTER 5

Thermal Equilibrium: System of Identical Point Molecules

5.1. Canonical Distribution

We are interested in a system of identical particles.
We proceed from the canonical distribution:

$$D = \frac{1}{\xi} e^{-\beta H}.$$

H is the sum of two terms:

$$H_p = \frac{1}{2m} (p_1^2 + p_2^2 + \cdots + p_N^2),$$

$$H_C = \sum_J V_J + \tfrac{1}{2} \sum_J \sum_K W_{JK}.$$

We can therefore put D in the form of a product:

$$D = D_p D_C. \tag{1}$$

It is convenient to assume that each factor is normalized separately.
We then have

$$D_p = \frac{1}{\xi_p} e^{-\beta H_p}, \tag{2}$$

$$D_C = \frac{1}{\xi_C} e^{-\beta H_C}. \tag{3}$$

The norm of D_p can be calculated explicitly. The result is the classical one:

$$\xi_p = (2\pi m k T)^{3/2N}. \tag{4}$$

The densities can be calculated from the formulae in Chapter 1:

$$n_1 = N \int D_C \, d^3 r_2 \, d^3 r_3 \ldots d^3 r_N, \tag{5}$$

$$n_{12} = N(N-1) \int D_C \, d^3 r_3 \, d^3 r_4 \ldots d^3 r_N,$$

$$\ldots$$

Let us introduce the Maxwell function:

$$f_i = \frac{1}{(2\pi mkT)^{3/2}} \exp\left(-\beta \frac{p_i^2}{2m}\right), \tag{6}$$

which satisfies the normalization condition

$$\int f_i \, d^3 p_i = 1.$$

The simple, double and triple densities in phase can be expressed by means of this function and the ordinary densities:

$$\mu_1 = n_1 f_1, \tag{7}$$
$$\mu_{12} = n_{12} f_1 f_2,$$
$$\mu_{123} = n_{123} f_1 f_2 f_3,$$
$$\cdots$$

There is no correlation between the momenta—or the velocities—and the spatial coordinates. There is no correlation between the velocities of two different molecules or between the velocity components of the same molecule along different axes. Maxwell's law expressed by the formula (6) is clearly fundamental.

Any local average of a function of the velocity is independent of the space coordinates. We recall, for example, that

$$\langle p_1^4 \rangle = \int p_1^4 f_1 \, d^3 p_1.$$

All the odd functions of the momentum have a zero average

$$\langle p_1 \rangle = \langle p_1 q_1 \rangle = \langle p_1^2 q_1 \rangle = \ldots = 0.$$

The velocity has an isotropic distribution law. In particular

$$\langle p_1^2 \rangle = \langle q_1^2 \rangle = \langle r_1^2 \rangle = mkT. \tag{8}$$

The kinetic pressure is diagonal and its three non-zero components are equal:

$$k_{1xx} = k_{1yy} = k_{1zz} = n_1 kT. \tag{9}$$

Similarly, the average kinetic energy does not depend on the molecular mass:

$$U_p = \tfrac{3}{2} kT \int n_1 \, d^3 r_1 = \tfrac{3}{2} NkT. \tag{10}$$

The heat flux vector, its structure and its parity being what they are, is identically zero.

We have recourse either to direct measurements or indirect measurements to verify Maxwell's law experimentally. Direct measurements consist of allowing particles to filter through an orifice and measuring the velocities with a rotating selector. Some measurements have been made on ordinary gases but extensive measurements have been made on neutrons. In an atomic pile equipped with a moderator (water, heavy water or graphite as the case may be) the neutrons live long enough to be "thermalized". Interpretation of the experiments must take incomplete thermalization into consideration: on the one hand there is an excess of fast neutrons and on the other hand there are some alterations due to diffusion through the reflector.

Problems relating to configuration space are by no means as simple. They are dominated by the formulae (5) which are more concise than practical. New recurrence relations will let us connect the simple density and the multiple densities with each other. The simplest thing to do is to use the chain of recurrence equations of the second kind:

$$\left(\frac{\partial}{\partial t} + \frac{p_1}{m}\frac{\partial}{\partial x_1} + X_1\frac{\partial}{\partial p_1}\right)\mu_1 + \int X_{12}\frac{\partial \mu_{12}}{\partial p_1}\,d^3r_2\,d^3p_2 = 0,$$

$$\left[\frac{\partial}{\partial t} + \frac{p_1}{m}\frac{\partial}{\partial x_1} + \frac{p_2}{m}\frac{\partial}{\partial x_2} + (X_1 + X_{12})\frac{\partial}{\partial p_2} + (X_2 + X_{12})\frac{\partial}{\partial p_2}\right]$$
$$\times \mu_{12} + \int\left(X_{13}\frac{\partial}{\partial p_1} + X_{23}\frac{\partial}{\partial p_2}\right)\mu_{123}\,d^3r_3\,d^3p_3 = 0,$$

...

and substitute equations (7). Here the time derivatives are zero. We thus obtain the recurrence equations of the second kind for the equilibrium:

$$kT(\partial n_1/\partial r_1) = F_1 n_1 + \int F_{12} n_{12}\,d^3r_2, \tag{11}$$

$$kT(\partial n_{12}/\partial r_1) = (F_1 + F_{12})\,n_{12} + \int F_{13} n_{123}\,d^3r_3, \tag{12}$$

$$kT(\partial n_{12}/\partial r_2) = (F_2 + F_{21})\,n_{12} + \int F_{23} n_{123}\,d^3r_3, \tag{13}$$

$$kT(\partial n_{123}/\partial r_1) = (F_1 + F_{12} + F_{13})\,n_{123} + \int F_{14} n_{1234}\,d^3r_4,$$

... (14)

This system is applicable whether the forces are long-range or short-range.

87

5.2. Grand Canonical Distribution and Partition Function

Here Maxwell's law and its consequences hold, as well as formulae (7) of section 1 and the equations of recurrence of the second kind. On the other hand, formulae (5) of section 1 require a special study which will be the subject of the next section. It is the compatibility of the grand canonical distribution and of the molecular disorder (see section 2.8) which makes us adopt this distribution in preference to any other. We shall be arguing within this framework henceforth.

The study of the properties of the grand partition function involves the calculation of its infinitesimal variations, both as a function of α and β and as a functional of the applied potential energy. Let us examine for the moment the effect of the variations of α and of β. We recall the following formulae:

$$\Xi = \sum_N \int \Gamma_N^{-1} \exp(\alpha N - \beta H_N) \, d\Omega_N, \tag{1}$$

$$D_N = \Gamma_N^{-1} \exp(\alpha N - \beta H_N - \ln \Xi). \tag{2}$$

By differentiation we derive from this

$$\delta \ln \Xi = \sum_N \int (N \, \delta\alpha - H_N \, \delta\beta) \, D_N \, d\Omega_N \tag{3}$$

and

$$\delta D_N = (N \, \delta\alpha - H_N \, \delta\beta - \delta \ln \Xi) \, D_N. \tag{4}$$

Formula (3) leads to the following results:

$$\partial \ln \Xi / \partial\alpha = \langle N \rangle, \tag{5}$$

$$\partial \ln \Xi / \partial\beta = -U, \tag{6}$$

where we see the occupation and the internal energy appear. We then have

$$\delta^2 \ln \Xi = \sum_N \int (N \, \delta\alpha - H_N \, \delta\beta) \, \delta D_N \, d\Omega_N, \tag{7}$$

or, bearing (4) in mind,

$$\delta^2 \ln \Xi = \sum_N \int [(N - \langle N \rangle) \, \delta\alpha - (H_N - U) \, \delta\beta]^2 \, D_N \, d\Omega_N, \tag{8}$$

which gives

$$\partial^2 \ln \Xi / \partial\alpha^2 = \langle \Delta N^2 \rangle = \delta \langle N \rangle / \partial\alpha, \tag{9}$$

$$\partial^2 \ln \Xi / \partial\beta^2 = \langle \Delta H^2 \rangle = -\partial U / \partial\beta, \tag{10}$$

$$\partial^2 \ln \Xi / \partial\alpha \, \partial\beta = -\langle \Delta N . \Delta H \rangle$$
$$= \partial \langle N \rangle / \partial\beta = -\partial U / \partial\alpha. \tag{11}$$

As the squared averages must be positive the result of the above formulae is that we have:

$$\frac{\partial \langle N \rangle}{\partial \alpha} \geqq 0, \qquad -\frac{\partial U}{\partial \beta} \quad \text{or} \quad \frac{\partial U}{\partial T} \geqq 0. \tag{12}$$

In addition the following quantity (x is arbitrary),

$$x^2 \langle \Delta H^2 \rangle + 2x \langle \Delta N \, \Delta H \rangle + \langle \Delta N^2 \rangle,$$

cannot be negative, which introduces a new inequality

$$\langle \Delta N^2 \rangle \langle \Delta H^2 \rangle \geqq \langle \Delta N \, \Delta H \rangle^2 \tag{13}$$

or

$$\left(-\frac{\partial U}{\partial \beta} \right) \frac{\partial \langle N \rangle}{\partial \alpha} \geqq \left(\frac{\partial U}{\partial \alpha} \right)^2. \tag{14}$$

5.3. Densities as a Function of α

The recurrence relations of the first kind

$$N = \int n_1 \, d^3 r_1, \tag{1}$$

$$(N - 1) n_1 = \int n_{12} \, d^3 r_2, \tag{2}$$

$$(N - 2) n_{12} = \int n_{123} \, d^3 r_3, \tag{3}$$

can be applied to the case when the number of particles is fixed and particularly to the canonical distribution. They cannot be transcribed as simply as the majority of the other formulae of the same kind when we pass to the grand canonical distribution. The results which can be obtained form a new chain of recurrence equations— the recurrence equations of the third kind.

Let us deal explicitly with the case of equations (1) and of (2). We must now bring in the densities with the suffix N. We rewrite:

$$\pi_N N = \int n_{N^1} \, d^3 r_1, \tag{4}$$

$$(N - 1) n_{N^1} = \int n_{N^{12}} \, d^3 r_2. \tag{5}$$

Let us sum over the suffix N. Equation (4) becomes

$$\langle N \rangle = \int n_1 \, d^3 r_1, \tag{6}$$

and equation (5) becomes

$$\sum_N N n_{N^1} = n_1 + \int n_{12} \, d^3 r_2. \tag{7}$$

But since

$$n_{N^1} = \frac{e^{\alpha N}}{\Xi} \times \text{quantity independent of } \alpha,$$

we obtain by taking the derivative

$$\partial n_{N^1}/\partial\alpha = \left(N - \frac{\partial \ln \Xi}{\partial\alpha}\right) n_{N^1}, \tag{8}$$

or

$$\partial n_{N1}/\partial\alpha = (N - \langle N \rangle)\, n_{N1}. \tag{9}$$

By substituting in equation (7) we obtain

$$\partial n_1/\partial\alpha = n_1 + \int (n_{12} - n_1 n_2)\, d^3 r_2. \tag{10}$$

This result can be generalized and gives the following system:

$$\langle N \rangle = \int n_1\, d^3 r_1, \tag{11}$$

$$\partial n_1/\partial\alpha = n_1 + \int (n_{21} - n_1 n_2)\, d^3 r_2, \tag{12}$$

$$\partial n_{12}/\partial\alpha = 2n_{12} + \int (n_{123} - n_{12} n_3)\, d^3 r_3, \tag{13}$$

$$\partial n_{123}/\partial\alpha = 3n_{123} + \int (n_{1234} - n_{123} n_4)\, d^3 r_4. \tag{14}$$

When molecular disorder obtains the integrations are here only over small regions.

Integration of equation (12) term by term over space leads to the following relation:

$$\partial \langle N \rangle/\partial\alpha = \int n_1\, d^3 r_1 + \int (n_{12} - n_1 n_2)\, d^3 r_1\, d^3 r_2, \tag{15}$$

or, as a consequence,

$$\langle \Delta N^2 \rangle = \int n_1\, d^3 r_1 + \int (n_{12} - n_1 n_2)\, d^3 r_1\, d^3 r_2. \tag{16}$$

This result has already been established in Chapter 2 and is a fundamental formula of the theory of fluctuations. We recall that it is equally valid for a partial volume provided that the integrations are limited to this volume.

5.4. Applied Energy Variations

We have studied the variation of the logarithm of the grand partition function for infinitesimal variations of α and of β. We still have to study the effect of the variations of the applied energy.

We return to formulae (1) and (2) of section 5.2 which now give

$$\delta \ln \Xi = -\beta \sum_N \int \left(\sum_1^N \delta V_J \right) D_N \, d\Omega_N, \tag{1}$$

or, because of the symmetry,

$$\delta \ln \Xi = -\beta \sum_N \int N \delta V_1 \, D_N \, d\Omega_N, \tag{2}$$

or finally

$$\delta \ln \Xi = -\beta \int n_1 \, \delta V_1 \, d^3 r_1. \tag{3}$$

The functional derivative of $\ln \Xi$ with respect to the applied energy is therefore the density (apart from a factor $-\beta$).

We shall calculate now the variation of n_1, proceeding from the expression

$$n_1 = \sum_N \int N D_N \, d^3 p_1 \, d\Omega_{2\ldots N}. \tag{4}$$

We obtain

$$\delta n_1 = \sum_N \int N \left[-\beta \left(\sum_{J=1}^N \delta V_J \right) - \delta \ln \Xi \right] D_N \, d^3 p_1 \, d\Omega_{2\ldots N}. \tag{5}$$

Here in the sum over J we must distinguish on the one hand the term with suffix 1 which plays a separate part, and on the other hand the suffices 2, 3, ..., N which all play the same part. Equation (5) can be rewritten as

$$\delta n_1 = -\beta \sum_N \int [N \delta V_1 + N(N-1) \delta V_2] D_N \, d^3 p_1 \, d\Omega_{2\ldots N} - n_1 \delta \ln \Xi. \tag{6}$$

We take $\delta \ln \Xi$ from equation (3) but replace the suffix 1 by the suffix 2 which gives the desired result:

$$\delta n_1 = -\beta \left[n_1 \delta V_1 + \int (n_{12} - n_1 n_2) \delta V_2 \, d^3 r_2 \right]. \tag{7}$$

This relation allows us to show that the second variation of the logarithm of the grand partition function is necessarily positive. It is equal in fact to the variance of $\sum_J \delta V_J$ multiplied by the square of β.

The arguments presented here can be extended to the case of a variation of the intermolecular energy.

5.5. h_1, h_{12}, ...

These quantities were introduced in section 5 of Chapter 2 in connexion with an analysis of the correlations. In the grand canonical distribution, and when we assume that there are only binary forces, their expression is simple:

$$h_1 = \alpha - \beta(p_1^2/2m) - \beta V_1,$$
$$h_{12} = -\beta W_{12},$$
$$h_{123} = h_{1234} = \cdots = 0.$$

5.6. Recurrence Equations of the Second Kind and Thermalization

Let us introduce into equations (11), (12) of section 5.1 a notation which allows us to express the correlations in terms of dimensionless quantities:

$$n_{12} = n_1 n_2 a_{12},$$
$$n_{123} = n_1 n_2 n_3 a_{12} a_{13} a_{23} a_{123}.$$

We obtain

$$kT(\partial \ln n_1/\partial r_1) = F_1 + \int F_{12} a_{12} n_2 \, d^3 r_2, \tag{1}$$

$$kT(\partial \ln a_{12}/\partial r_1) = F_{12} + \int F_{13} a_{13}(a_{23} a_{123} - 1) n_3 \, d^3 r_3, \tag{2}$$

$$kT(\partial \ln a_{123}/\partial r_1) = \int F_{14} a_{14}(a_{24} a_{34} a_{124} a_{134} a_{234} a_{1234} - a_{24} a_{124}$$
$$- a_{34} a_{134} + 1) n_4 \, d^3 r_4. \tag{3}$$
...

This result separates the problem of the configuration into two parts:

(1) on the one hand, the study of the correlations when the simple density, the temperature and the intermolecular energy are given, a problem which comes from equations (2), (3), ...;

(2) on the other hand, and secondly, assuming the problem of the correlations solved, the study of the relations between the simple density and the applied field. It is obvious that the reasonable way of studying this problem consists in trying to express the applied field as a function of the density rather than the opposite, the result being much more direct.

The results can be expressed with the help of expansions in rising powers of the density which can be obtained by successive approximations. The structure of the equations allows us to predict that the convergence will be better if molecular disorder obtains, because then the integrands which figure in (2), (3) are non-zero only in reduced domains. We should say at once that this is not exactly the method which we shall use, for practical reasons, firstly to avoid certain complex integrations connected with manipulating the system (1), (2), and also to profit from the results which we obtained when studying non-equilibrium system—results which can be applied just as well to equilibrium. But the path we shall follow and which tends to save the strength of the reader and the author, in no way contradicts the validity of the ideas that we have just stated in connexion with our recurrence equations. This is why it is important to note what has had to be assumed to establish these equations:

(1) the occupation is undetermined;
(2) there are no correlations between the configuration and the velocities;
(3) there are no correlations between the velocities of two different particles;
(4) the law giving the velocity distribution of a particle is given by Maxwell's law.

Conditions 2, 3 and 4 are not equivalent to the statement of the law governing the grand canonical distribution. They are broader. As a result the solution of the recurrence equations of the second kind can produce solutions which do not conform to thermodynamic equilibrium. This will definitely happen but these solutions should not be despised. They will correspond to metastable states. We shall say of a medium that satisfies the above conditions that it is "thermalized".

Regression operations carried out member by member on the system (11), (12) of section 5.1 lead to fresh relations. The latter are fairly simple because of the symmetry and the conditions at the limits.

A complete regression leads to the following relations:

$$\int F_1 n_1 \, d^3 r_1 = \int F_1 n_{12} \, d^3 r_1 \, d^3 r_2$$
$$= \int F_1 n_{123} \, d^3 r_1 \, d^3 r_2 \, d^3 r_3 = \cdots = 0. \qquad (4)$$

The first of these results expresses the fact that the average applied force is zero, which is obviously for a system at rest on the average.

Let us multiply equation (13) of section 5.1 term by term by d^3r_2 and integrate. We get

$$\int F_2 n_{12} \, d^3r_2 = \int F_{12} n_{12} \, d^3r_2. \tag{5}$$

This relation allows us to eliminate the integral of the intermolecular forces in (11) of section 5.1. We have

$$kT(\partial n_1/\partial r_1) = F_1 n_1 + \int F_2 n_{12} \, d^3r_2, \tag{6}$$

or, allowing for (4),

$$kT(\partial n_1/\partial r_1) = F_1 n_1 + \int F_2 (n_{12} - n_1 n_2) \, d^3r_2. \tag{7}$$

In the molecular disorder hypothesis the integral which figures in this formula affects only a small region around the point r_1. Formula (7) just allows us to bring out the cases when the molecular disorder hypothesis is indefensible.

A first example is that when the applied field is uniform in the region in question. We assume that it runs along the vertical Oz. Assuming that the correlations fade out rapidly with distance, we obtain

$$\frac{\partial n_1}{\partial x_1} = 0, \qquad \frac{\partial n_1}{\partial y_1} = 0,$$

$$kT \frac{\partial n_1}{\partial z_1} = Z \left[n_1 + \int (n_{12} - n_1 n_2) \, d^3r_2 \right]. \tag{8}$$

These results mean first of all that the surfaces of equal density are horizontal planes. Observation of a liquid in equilibrium with its vapour in the field of gravity shows us that this result is correct except in the vicinity of the walls.

Experience also tells us that the transition from a liquid to its vapour is achieved in a very thin layer—we are ignoring the critical region here. In this laminar zone the density varies with very great rapidity along the normal. In regions where the laminar zone is curved, therefore, the relations (8) contradict observation, and it is not a question of a microscopic phenomenon since the meniscus phenomenon is visible to the naked eye. It is therefore necessary to assume that there are correlations between the points of the laminar zone and the adjacent points of the walls. These correlations should have enough effect to balance the density gradient which may be very great. This does not necessarily mean that these correlations are very strong since calculation of the integral

$$\int F_2(n_{12} - n_1 n_2)\, \mathrm{d}^3 r_2$$

does not only introduce the correlations but also the forces at the wall in particular.

Studying the behaviour of a fluid at the wall is as difficult as the problem of the laminar zone itself. It does not perhaps deserve very deep treatment since the wall that we have imagined for retaining the fluid is very schematic. It is not a molecular wall. In any case we can arbitrarily ascribe to it a highly accentuated grain or roughness. The force field which represents it and is essentially repulsive can have an attractive component. This complex structure does not alter the validity of our arguments and would possibly enable us to face various experimental situations and the various ways in which a liquid can be attached to a wall.

Therefore we still have to make a detailed analysis of what the correlations are.

We notice that problems connected with condensation can only become worse when

$$Z = 0,$$

i.e. in the absence of gravity. These problems will become quite acute with the development of space research (NASA, 1963; in this reference experiments on free fall in a tower are discussed).

5.7. Effects of a Variation of the Volume

In section 5.4 the nature of the variations of the applied energy was not specified. They may be variations in the interior of the given volume or they may equally be variations of this volume itself since its definition is incorporated in the force field.

Treating the variations of the volume by the procedure of section 5.4 is a little suspect since we are now varying an infinite quantity. This, however, is only a mathematical difficulty which can be overcome by appropriate arguments; we shall neglect the intermediate steps of the rigorous treatment.

We shall imagine an expansion of all the system's dimensions along the x-axis. The applied potential energy which was

$$V_J(x, y, z)$$

becomes

$$V_J((1 - \lambda)\, x, y, z), \tag{1}$$

λ being a small positive quantity. We should make it clear that this change is properly speaking equivalent to an increase in volume only if the only part that V_J has to play is to represent the wall. If there is likewise a field distributed throughout the whole of the given volume the latter is also altered by the perturbation we have just defined.

Anyhow, without further specifying whether the field is limited or not to the wall, formula (1) gives the following variation:†

$$\delta V_J = \lambda x_J X_J. \tag{2}$$

Formula (3) of section 5.4 can now be written as

$$\delta \ln \varXi = -\beta \int n_1 x_1 X_1 \, d^3 r_1. \tag{3}$$

We multiply the first recurrence equation of the second kind term by term by x_1 and obtain

$$n_1 x_1 X_1 = kT x_1 \frac{\partial n_1}{\partial x_1} - \int x_1 X_{12} n_{12} \, d^3 r_2, \tag{4}$$

then, by integrating in accordance with the procedure which leads to equation (27) of section 1.6,

$$\int n_1 x_1 X_1 \, d^3 r_1 = - \int kT n_1 \, d^3 r_1 + \tfrac{1}{2} \int (x_2 - x_1) X_{12} n_{12} \, d^3 r_1 \, d^3 r_2. \tag{5}$$

We therefore have

$$\delta \ln \varXi = \lambda \beta \int P_{xx} \, d^3 r. \tag{6}$$

A uniform expansion by a factor $1 + \lambda$ similarly leads to a variation

$$\delta \ln \varXi = \lambda \beta \int (P_{xx} + P_{yy} + P_{zz}) \, d^3 r. \tag{7}$$

If we assume that the medium is uniform and isotropic before the perturbation—we shall return to this kind of situation—we finally obtain the derivative of the logarithm with respect to the volume

$$\partial \ln \varXi / \partial V = \beta P. \tag{8}$$

We shall return to formulae (7) and (8) in the following chapters.

---

† In the remainder of this section, we do not imply summation over "repeated indices" (see Note on Notation on p. xix); xX, $x\partial n/\partial x$, ... therefore indicate just *one* term.

CHAPTER 6

Fine Structure in Thermal Equilibrium

6.1. Fine Structure Equation

In a medium in equilibrium the correlations relate only to space. We propose to express these correlations as a function of the density and the intermolecular force. We are dealing with the case when the occupation is poorly defined.

There are several ways of solving this problem: (1) making direct use of the grand canonical formalism which proceeds from the densities of occupation; (2) making use of the recurrence equations of the second kind; (3) make use of equations (6) of section 3.8 which express h_{12}, h_{123}, ... as functions of the correlations. The first method used by the majority of authors is uselessly restrictive; it is true that it is the one that lends itself best to systematic development (Mayer and Mayer, 1940). The second is complicated but it has the advantage of only assuming that the medium is thermalized. The third is based like the first on the grand canonical distribution but it has the advantage of roughly following the same development stages as the second method and it is this third method that we shall use here. Bearing in mind the expressions for the h which we made explicit at the end of the last chapter, equations (6) of section 3.8 which we need become

$$
\begin{aligned}
- \beta W_{12} = \ln a_{12} &- \int n_3 (a_{13} a_{23} a_{123} - a_{13} - a_{23} + 1) \, \mathrm{d}^3 r_3 \\
&+ \tfrac{1}{2} \int n_3 n_4 (a_{13} a_{14} a_{23} a_{24} a_{34} a_{123} a_{124} a_{134} a_{234} a_{1234} \\
&- a_{13} a_{23} a_{123} a_{14} a_{24} a_{124} - a_{13} a_{14} a_{34} a_{134} \\
&- a_{23} a_{24} a_{34} a_{234} + a_{13} a_{14} + a_{23} a_{24} + a_{34} - 1) \\
&\times \mathrm{d}^3 r_3 \, \mathrm{d}^3 r_4 - \cdots ,
\end{aligned}
\tag{1}
$$

97

$$0 = \ln a_{123} - \int n_4(a_{14}a_{24}a_{34}a_{124}a_{134}a_{234}a_{1234} - a_{14}a_{24}a_{124}$$
$$- a_{14}a_{34}a_{134} - a_{24}a_{34}a_{234} + a_{14} + a_{24} + a_{34} - 1)$$
$$\times \, d^3r_4 + \cdots,$$
$$0 = \ln a_{1234} - \cdots.$$

Let us solve them by successive approximations, the density being considered a first-order quantity. In zeroth order we have

$$a_{12} = \exp\left(-\beta W_{12}\right);$$
$$a_{123} = a_{1234} = \cdots = 1. \tag{2}$$

It should be noted at once that this kind of approximation will be unsuitable for the case of Coulomb forces because it introduces for a_{12} a decrease with distance that is too slow. We shall leave the case of the long-range forces alone. We shall say

$$g_{12} = \exp\left(-\beta W_{12}\right) - 1, \tag{3}$$

while observing that g_{12} decreases rapidly when the distance increases.

Up to second order we have

$$\ln a_{12} = -\beta W_{12} + \int n_3 g_{13} g_{23} \, d^3r_3,$$
$$\ln a_{123} = \int n_4 g_{14} g_{24} g_{34} \, d^3r_4,$$
$$\ln a_{1234} = \int n_5 g_{15} g_{25} g_{35} g_{45} \, d^3r_5. \tag{4}$$

By virtue of the short range of g the integrations cover only small regions of space in the vicinity of the fixed points. In addition the asymptotic properties of the a's are of the "molecular disorder" type. The same will be the case further on.

In the same approximation equations (4) can be rewritten as

$$a_{12} = \exp\left(-\beta W_{12}\right)\left(1 + \int n_3 g_{13} g_{23} \, d^3r_3\right),$$
$$\varepsilon_{123} = a_{123} - 1 = \int n_4 g_{14} g_{24} g_{34} \, d^3r_4,$$
$$\varepsilon_{1234} = a_{1234} - 1 = \int n_5 g_{15} g_{25} g_{35} g_{45} \, d^3r_5. \tag{5}$$

To obtain the next approximation of a_{12} we return to the firs
equation (1).

In the first integral we substitute for the a their expressions (5) and in the second their expressions (2). We obtain

$$\int n_3(a_{13}a_{23}a_{123} - a_{13} - a_{23} + 1)\, d^3r_3$$

$$= \int n_3(a_{13}a_{23}\varepsilon_{123} + \varepsilon_{13}\varepsilon_{23})\, d^3r_3 \simeq \int n_3 g_{13}g_{23}\, d^3r_3$$

$$+ \int n_3 \left[(1 + g_{13})(1 + g_{23})\int n_4 g_{14}g_{24}g_{34}\, d^3r_4 + g_{13}(1 + g_{23})\right.$$

$$\left. \times \int n_4 g_{24}g_{34}\, d^3r_4 + g_{23}(1 + g_{13})\int n_4 g_{14}g_{34}\, d^3r_4\right] d^3r_3, \quad (6)$$

$$\tfrac{1}{2}\int n_3 n_4(\ldots)\, d^3r_3\, d^3r_4$$

$$\simeq \tfrac{1}{2}\int n_3 n_4(a_{13}a_{14}a_{23}a_{24}a_{34} - a_{13}a_{14}a_{23}a_{24} - a_{13}a_{14}a_{34}$$

$$- a_{23}a_{24}a_{34} + a_{13}a_{14} + a_{23}a_{24} + a_{34} - 1)\, d^3r_3\, d^3r_4$$

$$\simeq \tfrac{1}{2}\int n_3 n_4 g_{34}(g_{13}g_{14}g_{23}g_{24} + g_{13}g_{14}g_{23} + g_{13}g_{14}g_{24}$$

$$+ g_{13}g_{23}g_{24} + g_{14}g_{23}g_{24} + g_{13}g_{23} + g_{13}g_{24} + g_{14}g_{23}$$

$$+ g_{14}g_{24})\, d^3r_3\, d^3r_4. \quad (7)$$

This kind of notation is rather hard to read. A representation using diagrams helps the work. The following diagrams are sufficiently explicit for the principle of the method:

$$g_{12} = 1 - 2,$$

$$g_{13}g_{23} = \tfrac{1}{2}{>}3,$$

$$g_{13}g_{14}g_{23} = \tfrac{1}{2}{\lessgtr}\tfrac{4}{3}.$$

With these diagrams our two integrals (6) and (7) can be written respectively as

$$\int n_3 \tfrac{1}{2}{>}\, d^3r_3 + \int n_3 n_4 \left(\tfrac{1}{2}{\succ} + 2 \times \tfrac{1}{2}\boxminus + 2 \times \tfrac{1}{2}\boxslash + \tfrac{1}{2}\boxtimes + 2 \times \tfrac{1}{2}\boxbslash\right)$$

$$\times d^3r_3\, d^3r_4, \quad (8)$$

$$\tfrac{1}{2}\int n_3 n_4 \left(\tfrac{1}{2}\boxtimes + 2 \times \tfrac{1}{2}\boxminus + 2 \times \tfrac{1}{2}\boxslash + 2 \times \tfrac{1}{2}{\succ} + 2 \times \tfrac{1}{2}\boxbslash\right) d^3r_3\, d^3r_4. \quad (9)$$

There is no point in making the dummy suffices explicit and this allows certain parts to be combined. We finally obtain, up to third order,

$$\ln a_{12} = -\beta W_{12} + \int \tfrac{1}{2}{>}\, n_3\, d^3r_3 + \int \left(\tfrac{1}{2}\tfrac{1}{2}\boxtimes + \tfrac{1}{2}\boxminus + \tfrac{1}{2}\boxslash + \tfrac{1}{2}\boxbslash\right)$$

$$\times n_3 n_4\, d^3r_3\, d^3r_4 + \cdots. \quad (10)$$

99

This is the equation of the fine structure. It is noticeable that it does not contain the applied field. Its generalization to higher-order terms will not be examined here. But it is not hard to state the rules for writing them:

1. We meet successively with diagrams with 3, 4, 5, ... vertices.

2. There is no vertex which itself joins two elements of the diagram. Forbidden examples:

3. 1 and 2 are never directly connected.

4. Starting from 1 and 2 there are never two independent diagrams. Forbidden example:

5. All the diagrams are preceded by a plus sign. Numerical factors appear:

If there are 2 equivalent vertices, a factor of $1/2!$,

If there are 3 equivalent vertices, a factor of $1/3!$,

If there are 4 equivalent vertices, a factor of $1/4!$,

If there are two sets of two equivalent vertices, a factor of $(1/2!)^2$ and so on. More complete rules will be given in section 6.3.

In the case of long-range forces, particularly in the case of Coulomb forces and of completely ionized media, it is more appropriate, so as to get to grips with the problem of the correlations, to use the recurrence equations of the second kind which permit a fairly physical approach to the question (Delcroix, 1963).

On the other hand, it should be pointed out that certain relations similar to the ones above can be obtained not by expressing the correlations as functions of the density and the molecular forces, but by expressing the force and higher-order correlations as functions of the density and of a_{12}.

6.2. Grand Partition Function

The cumulant method made it possible to obtain [formula (9) of section 3.6] an appropriate expression for $D_0 = \Xi^{-1}$ which we can rewrite in the case of equilibrium as

$$\ln \Xi = \int n_1 \, d^3 r_1 - \frac{1}{2} \int n_1 n_2 (a_{12} - 1) \, d^3 r_1 \, d^3 r_2$$

$$+ \frac{1}{6} \int n_1 n_2 n_3 (a_{12} a_{13} a_{23} a_{123} - a_{12} - a_{13} - a_{23} + 2)$$

$$\times d^3 r_1 \, d^3 r_2 \, d^3 r_3 - \frac{1}{24} \int n_1 n_2 n_3 n_4 (a_{12} a_{13} a_{14} a_{23} a_{24} a_{34}$$

$$\times a_{123} a_{124} a_{134} a_{234} a_{1234} - a_{12} a_{13} a_{23} a_{123} - a_{12} a_{14} a_{24} a_{124}$$

$$- a_{13} a_{14} a_{34} a_{134} - a_{23} a_{24} a_{34} a_{234} - a_{12} a_{34} - a_{13} a_{24}$$

$$- a_{14} a_{23} + 2 a_{12} + 2 a_{13} + 2 a_{14} + 2 a_{23} + 2 a_{24}$$

$$+ 2 a_{34} - 6) \, d^3 r_1 \, d^3 r_2 \, d^3 r_3 \, d^3 r_4 + \cdots. \tag{1}$$

As it is a_{12} we need and not its logarithm we rewrite the fine structure equation as follows:†

$$a_{12} = (1 + g_{12}) \left[1 + \int \tfrac{1}{2} {>} \; n_3 \, d^3 r_3 \right.$$

$$\left. + \int \left(\tfrac{1}{2} \tfrac{1}{2} \boxtimes + \tfrac{1}{2} \boxed{} + \tfrac{1}{2} \boxed{} + \tfrac{1}{2} \diamondsuit + \tfrac{1}{2} \boxed{} \right) n_3 n_4 \, d^3 r_3 \, d^3 r_4 + \cdots \right]. \tag{2}$$

This equation can be used to eliminate a_{12}. The corresponding equations of the preceding section can likewise be used to eliminate a_{123}, a_{1234}, ...

After making all the calculations we obtain

$$\ln \Xi = \int n_1 \, d^3 r_1 - \int \frac{1}{2} n_1 n_2 g_{12} \, d^3 r_1 \, d^3 r_2$$

$$- 2 \int \frac{1}{6} n_1 n_2 n_3 \triangleleft \, d^3 r_1 \, d^3 r_2 \, d^3 r_3 - 3 \int n_1 n_2 n_3 n_4$$

$$\times \left(\frac{1}{24} \diamondsuit + \frac{1}{4} \diamondsuit + \frac{1}{8} \, \square \right) d^3 r_1 \, d^3 r_2 \, d^3 r_3 \, d^3 r_4 - \cdots. \tag{3}$$

† It will be noted that there are diagrams here that do not figure in (10) of section 6.1. The rules stated there apply only to the expansion of $\ln a_{12}$.

Fine Structure in Thermal Equilibrium

Here all the vertices correspond to dummy suffices. The diagrams are so compact that in the multiple integrals the integration domains are microscopic—1, 2, 3, 4, ... remain adjacent to each other when we insist that the integrand should not be zero—except for the last integration which covers the whole of the available region.

6.3. Density Distribution

Amongst our basic formulae we have not yet used the one which expresses h_1 as a function of the correlations, i.e. formula (6) of section 2.8:

$$h_1 = \ln \Gamma \mu_1 - \int \mu_2 (a_{12} - 1) \, d\Omega_2 + \cdots. \tag{1}$$

We could use this relation by following the same procedure of substitution as served us in the earlier sections. But now that we know $\ln \varXi$ we can proceed to shorter calculations.

In section 5.4 we calculated the variation of $\ln \varXi$ for small variations of the applied potential energy. The result is as follows:

$$\delta \ln \varXi = -\beta \int n_1 \, \delta V_1 \, d^3 r_1, \tag{2}$$

which shows that the functional derivative of $\ln \varXi$ with respect to the applied potential energy is

$$-\beta n_1.$$

Let us now calculate the variation of $\ln \varXi$ using formula (3) from the last section. We obtain

$$\delta \ln \varXi = \int \delta n_1 \, d^3 r_1 - \tfrac{1}{2} \int (\delta n_1 n_2 + n_1 \, \delta n_2) \, g_{12} \, d^3 r_1 \, d^3 r_2$$

$$- \frac{2}{6} \int (\delta n_1 n_2 n_3 + n_1 \, \delta n_2 n_3 + n_1 n_2 \, \delta n_3)$$

$$\times \, g_{12} g_{13} g_{23} \, d^3 r_1 \, d^3 r_2 \, d^3 r_3 - \cdots, \tag{3}$$

which can be rewritten by taking advantage of the symmetry and eliminating δn_1 from the integrals except the first one

$$\delta \ln \varXi = \int \delta n_1 \, d^3 r_1 - \int n_1 \, \delta n_2 g_{12} \, d^3 r_1 \, d^3 r_2$$

$$- \tfrac{1}{2} \int n_1 (n_3 \, \delta n_2 + n_2 \, \delta n_3) \, g_{12} g_{13} g_{23} \, d^3 r_1 \, d^3 r_2 \, d^3 r_3 - \cdots.$$

Making (2) and (4) identical we obtain

$$-\beta \, \delta V_1 = \frac{1}{n_1} \delta n_1 - \int \delta n_2 g_{12} \, \mathrm{d}^3 r_2 - \frac{1}{2} \int (\delta n_1 n_3 + \delta n_3 n_2)$$
$$\times \, g_{12} g_{13} g_{23} \, \mathrm{d}^3 r_2 \, \mathrm{d}^3 r_3 - \cdots. \tag{4}$$

This functional relation can be integrated. The expansion (1) supplies the integration constant. The result is as follows:

$$\ln \frac{(2\pi m k T)^{3/2}}{\Gamma} + \alpha - \beta V_1$$
$$= \ln n_1 - \int g_{12} n_2 \, \mathrm{d}^3 r_2 - \frac{1}{2} \int {}^1\!\triangleleft \, n_2 n_3 \, \mathrm{d}^3 r_2 \, \mathrm{d}^3 r_3$$
$$- \int \left(\frac{1}{6} {}^1\!\diamondsuit + \frac{1}{2} {}^1\!\diamondsuit + \frac{1}{2} {}^1\!\diamondsuit + \frac{1}{2} {}^1\!\diamondsuit \right)$$
$$\times \, n_2 n_3 n_4 \, \mathrm{d}^3 r_2 \, \mathrm{d}^3 r_3 \, \mathrm{d}^3 r_4 - \cdots. \tag{5}$$

This equation allows us to calculate the distribution of the density in a given applied field. In the case when the density is low the second term is reduced to the logarithm of the density: the equation reduced in this way is known as the Laplace equation. We shall therefore give the complete equation the name of the generalized Laplace equation. We should not attempt to solve equation (5) by successive approximations, since the expansion thus obtained in increasing powers of $\exp(-\beta V_1)$, which could, moreover, be obtained much more directly from basic principles, is not convergent. The explanation of this is that a field does not act uniquely in a local manner on a fluid. The variations of the field in a certain region affect the whole density distribution. There are other methods which must be applied.

It will be noted that all that counts is the difference between the chemical potential and the applied potential energy, both of which are defined apart from additive constants.

The writing of the higher-order terms is based on rules similar to those which we have just stated for the fine structure, but a little simpler:

(1) no vertex by itself joins two elements of the diagram;
(2) all the diagrams are preceded by a minus sign;
(3) the factorials are determined by the symmetry as before, but we should make this clear: a diagram like those which figure in the generalized Laplace equation or in the equation of

the fine structure is an object which is characterized by points and connexions between these points. Certain points are numbered in advance (there is one such point, numbered 1, in the Laplace equation and two such points, numbered 1 and 2, in the equation of the fine structure). The other points are "dummies". Let us assume that they are N in number. Let us give them each a number running from 2 to $N + 2$ in the first case and from 3 to $N + 3$ in the second. Let us number the dummy vertices in all possible ways: the number of objects obtained is $P \leqq N!$ The numerical factor which should correspond to the particular diagram studied is $P/N!$

Examples $N = 4$:

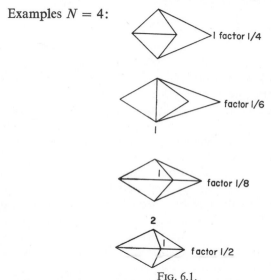

Fig. 6.1.

The fine structure equation and the generalized Laplace equation (apart from the significance of α) can be derived uniquely from the recurrence equations of the second kind. They are therefore applicable outside the grand canonical equilibrium: it is sufficient that the fluid should be thermalized.

6.4. The Case of Uniform Media

We should make it clear here that we are dealing only with the case of short-range forces, although a certain number of results can be generalized for electrically neutral ionized media.

In a medium of this kind the simple density does by definition not depend on the spatial coordinates. The result is that the force field is zero.

The uniform fluids that we are observing normally do not satisfy this condition rigorously. On the one hand, in the vicinity of the walls complex actions create certain density gradients. On the other hand, gravity involves a non-uniformity that is more or less accentuated. The upper layers press down on the lower layers which are denser. The less dense the medium is on the average and the less compressible it is, the weaker is this effect. Depending on the circumstances we shall have to take into consideration the heterogeneous features that we have just mentioned or neglect them.

Let us assume them to be negligible. Formula (3) of section 6.2 shows us that, except for certain superficial effects which will not come into play in large volumes, i.e. those of normal experiments, $\ln \varXi$ is proportional to the volume. This statement assumes convergence of the expansion (3) in the conditions that we have mentioned in this connexion.

The study of pressure will allow us to find a physical expression of the proportionality constant.

We recall that in a uniform medium the pressure is isotropic and that it is then given by formula (18) of section 1.6 completed by the kinetic pressure. We write again

$$ P = nkT - \frac{2\pi}{3} \int_0^\infty n_{12}(r) \frac{\mathrm{d}W}{\mathrm{d}r} r^3 \, \mathrm{d}r, $$

or, equally well,

$$ P = nkT - \frac{1}{2} \int n_{12}(x_2 - x_1) \frac{\partial W_{12}}{\partial x_2} \, \mathrm{d}^3 r_2. \tag{1} $$

By substituting for n_{12} in this formula its fine-structure expansion written for a uniform medium we should obtain an expansion of the pressure in a complete series of the density. To avoid this calculation we shall proceed to the following arguments:

Let us first consider a medium which is not particularly uniform or particularly isotropic. We rewrite the first recurrence equation:

$$ kT(\partial n_1/\partial r_1) = n_1 F_1 + \int n_{12} F_{12} \, \mathrm{d}^3 r_2, \tag{2} $$

and the formula which expresses the essential properties of the intermolecular pressure tensor:

$$ \int n_{12} X_{12} \, \mathrm{d}^3 r_2 = -\partial p_{1xy}/\partial y_1. \tag{3} $$

105

Here the kinetic pressure is a scalar. Its value is kTn_1. In any case we can put it in tensor form:

$$k_{1xy} = kTn_1\, \delta_{xy}.$$

This said, the total pressure satisfies the relation

$$n_1 X_1 = \partial P_{1xy}/\partial y_1. \tag{4}$$

This relation is the application of the hydrodynamic equation of motion to the static case. Let us assume that the applied field leaves the fluid almost uniform and at the same time is almost isotropic. The pressure is almost a scalar. In the above equation the non-uniformity is allowed for by the derivative. It is therefore legitimate to neglect the anisotropy, the cumulative effects of both being of second order. The pressure is therefore considered a scalar. We obtain

$$n_1 F_1 = \partial P_1/\partial r_1, \tag{5}$$

which is the elementary equation for the equilibrium of "liquids".

On the other hand, by deriving the generalized Laplace equation we obtain the following relation after integrations by parts which lead to the fact that the derivative in the second term acts only on the density:

$$n_1 F_1 = kT\big[(\partial n_1/\partial r_1) - n_1 \int g_{12}(\partial n_2/\partial r_2)\, \mathrm{d}^3 r_2$$

$$- n_1 \int g_{12} g_{13} g_{23} n_3 (\partial n_2/\partial r_2)\, \mathrm{d}^3 r_2\, \mathrm{d}^3 r_3 - \cdots\big]. \tag{6}$$

But since we must also neglect the effects of the second order of the non-uniformity, we can everywhere replace

$$\partial n_2/\partial x_2 \quad \text{by} \quad \partial n_1/\partial x_1,$$

$$n_3, n_4, \ldots \quad \text{by} \quad n_1.$$

We thus obtain, by combining (5) and (6),

$$\partial P/\partial n = kT\left(1 - n \int g_{12}\, \mathrm{d}^3 r_2 - n^2 \int g_{12} g_{13} g_{23}\, \mathrm{d}^3 r_2\, \mathrm{d}^3 r_3 - \cdots\right). \tag{7}$$

Or, by integrating,

$$P = kT\left(n - \tfrac{1}{2} n^2 \int g_{12}\, \mathrm{d}^3 r_2 - \tfrac{1}{3} n^3 \int g_{12} g_{13} g_{23}\, \mathrm{d}^3 r_2\, \mathrm{d}^3 r_3 - \cdots\right). \tag{8}$$

The pressure is zero when the density is zero: there is no integration constant. The equation obtained expresses the pressure as a function of the temperature and of the density, assuming the inter-molecular forces to be known. This is Ursell's equation of state. The arguments imply that we can pass continuously from the observed density to zero density, dealing at each step with a quasi-uniform medium. There is only one apparent difficulty there because the development that could be derived from equation (1) and the one we have just obtained correspond to one another term for term independently of the value of the density.

Combining it with (3) of section 6.2 gives us the following expression:

$$\ln \Xi = PV/kT, \qquad (9)$$

which completes a remark made above. This result is in accordance with formula (8) of section 5.7.

When the density is moderate the pressure is reduced to the kinetic pressure. Then

$$P = nkT. \qquad (10)$$

We had stated without justifying it the value of the Boltzmann constant at the time of defining the entropy. In fact, this constant is the quotient of the perfect gas constant divided by the Avogadro number:

$$k = R/N_0. \qquad (11)$$

Equation (10) is then formally equivalent to the equation of state of perfect gases. The temperature which we defined in equation (10) of section 4.3 is therefore the same as the temperature of the perfect gas thermometer, which justifies the definition in question and also the choice of the value of k.

To finish with the Ursell equation we rewrite it below with the help of diagrams of the second kind, making one more term explicit:

$$P = nkT \left[1 - \tfrac{1}{2} n \int 1 - \mathrm{d}^3 r_2 - \tfrac{1}{3} n^2 \int 1 \triangleleft \, \mathrm{d}^3 r_2 \, \mathrm{d}^3 r_3 \right.$$
$$\left. - \tfrac{3}{4} n^3 \int (\tfrac{1}{6} 1 \Leftrightarrow + 1 \Leftrightarrow + \tfrac{1}{2} 1 \Diamond) \, \mathrm{d}^3 r_2 \, \mathrm{d}^3 r_3 \, \mathrm{d}^3 r_4 - \cdots \right]. \qquad (12)$$

From the point of view of convergence this expansion of the pressure brings forth two observations.

The first is that, it being well understood that in all respects the forces are short-range forces, the range of the integrands themselves increases with the order of the terms. The high-order terms flood the whole of the available volume, so much so that it is no longer legitimate to leave the integration limits undefined: the integrals of a definite rank no longer have the same value depending on whether the point r_1 is relatively close to or on the contrary definitely a long way from the walls.

We shall avoid this difficulty by assuming that the high-order terms make only a negligible contribution to the result. If it were otherwise there would no longer exist physically uniform media.

This granted, it does not matter keeping all the terms in equatin (12) while looking upon the medium as infinite. As a complement of the last hypothesis we assume that then the series (12), a complete series with respect to the density, is convergent or that at the very least there is an analytical continuation of it which makes it usable throughout the region of densities that is physically of interest.

The other expansions that we have used are not complete series. They are functionals of the density. We shall formulate similar hypotheses for them to those above. The convergence is not necessarily of the same quality for all the expansions. It is linked to the structure of the diagrams and probably the more compact the diagrams the better it is.

Mathematically, these questions of convergence are far from being solved. Physically, they correspond to the fact that the properties of a fluid are defined at each of its points by the structure of the immediate neighbourhood: the molecules amalgamate with each other in small groups, then these groups amalgamate with each other until they build up the macroscopic medium, two distant groups having no other links between them than contacts from group to group.

There is thus an apparent disagreement between the physical reality and the general laws relating to thermal equilibrium which consider all the molecules at once. This is why, as we have been able to see in reading the preceding pages, it is not without effort that we can pass from those laws to a fine-grained description.

Let us return to the formula (9). It allows us to eliminate the grand partition function from formula (6) of section 4.4. We obtain

$$U + PV - TS = \omega\langle N \rangle. \tag{13}$$

Our set of macroscopic formulae, i.e. those which relate only to U, P, V, T, S, $\langle N \rangle$ and ω, is now equivalent to the classical formulae of thermodynamics. What we have called internal energy, pressure, temperature, entropy and chemical potential behave exactly like the classical quantities of the same name. The two disciplines complement each other harmoniously: one is more refined but more formal, the other is broader. Thermodynamics gains in precision when statistical physics casts light on the ideas of temperature and of entropy. Conversely, statistical research relating to entropy receives a certain encouragement when the above-mentioned equivalence allows them to take shelter under the wing of the Carnot principle. We still have to show explicitly that for reversible changes we can write

$$dQ = T\,dS.$$

To end this section we should stress that the generalized Laplace equation (provided we forget for the moment the significance of the parameter α), the equation of the fine structure and the Ursell pressure equation in fact require for their derivation only the thermalization hypothesis and that they do not necessarily imply the grand canonical equilibrium.

6.5. Uniform Medium Fluctuations

In studying the uniform fluid our independent variables are α, T and V. The functions of these quantities which we are to consider are $\langle N \rangle$, U, P, S and the fluctuations. For reasons of uniformity the quantities $\langle N \rangle$, U and S are proportional to the volume, whilst n and P depend only on α and on T. Thus by applying formulae (9) and (10) of section 5.2 we obtain

$$\frac{\langle \Delta N^2 \rangle}{\langle N \rangle^2} = \frac{1}{\langle N \rangle^2} \frac{\partial \langle N \rangle}{\partial \alpha}, \tag{1}$$

$$\frac{\langle \Delta H^2 \rangle}{U^2} = \frac{kT^2}{U^2} \frac{\partial U}{\partial T}. \tag{2}$$

These results show that the relative fluctuations approach zero when the volume increases. In the limit the grand canonical distribution behaves like the canonical distribution since the fluctuation of the number of particles present becomes negligible. And even like the microcanonical distribution since the energy fluctuations behave likewise. This is what justifies our considerations being devoted only to the grand canonical distribution. Examination of the case of a small system, for which the equivalence does not exist, will be carried out by immersing it in a larger system in order to be in the same convenient situation.

Equation (1) can take another form. In accordance with (5) of section 5.2 we have

$$\frac{\partial P}{\partial \alpha} = kTn. \tag{3}$$

Instead of expressing the pressure as a function of α let us take n as an independent variable. The above formula becomes

$$\frac{\partial P}{\partial n}\frac{\partial n}{\partial \alpha} = kTn, \tag{4}$$

or

$$\frac{1}{V}\frac{\partial \langle N \rangle}{\partial \alpha} = kTn \left(\frac{\partial P}{\partial n} \right)^{-1} = n^2 kT\chi, \tag{5}$$

χ being the isothermal compressibility. This quantity has the advantage of being open to experiment. Formula (1) will henceforth be written in the classical form:

$$\frac{\langle \Delta N^2 \rangle}{\langle N \rangle^2} = \frac{1}{\langle N \rangle} nkT\chi. \tag{6}$$

We have obtained in passing an expression for the derivative of the density with respect to α:

$$\frac{\partial n}{\partial \alpha} = n^2 kT\chi. \tag{7}$$

On the other hand, this derivative appears in the recurrence equations of the third kind. By combining the two expressions for our derivative we can derive an expression for the compressibility as a function of the densities:

$$n^2 kT\chi = n + \int (n_{12} - n_1 n_2)\, \mathrm{d}^3 r_2, \tag{8}$$

or, since the fluid is uniform and isotropic (Ornstein and Zernike, 1914),

$$n^2 kT\chi = n + 4\pi \int_0^\infty (n_{12}(r) - n^2)\, r^2 \,\mathrm{d}r. \tag{9}$$

The inequalities (12), (13) and (14) of section 5.2 which we have established in connexion with the second derivatives of $\ln \Xi$ can be transformed simply if we use the usual thermodynamic quantities.

The first of these inequalities at once becomes

$$\chi \geqq 0. \tag{10}$$

The third becomes

$$\chi C \geqq 0, \tag{11}$$

C being the specific heat at constant volume. The second gives

$$C \geqq 0 \tag{12}$$

when $\chi = 0$. Finally the theory of the fluctuations in a uniform medium gives two conditions:

$$\chi \geqq 0, \quad C \geqq 0. \tag{13}$$

It is not without interest to note that the condition (10) would be modified if the temperature were not positive.

6.6. Experimental Determination of the Fine-grained Structure

This determination can be made by measuring the angular distribution of monochromatic X-rays scattered by the uniform medium

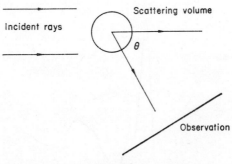

Fig. 6.2.

under study. Unless there is any correlation between the molecules the relative value of the scattered intensity without secondary effects would be

$$I = Nf^2,$$

N being the number of scattering atoms and f the form factor.

The correlations, which are assumed to be isotropic, modify the above result. The scattering is now given by the following formula

Fine Structure in Thermal Equilibrium

(Zernike and Prins, 1927):

$$I = Nf^2 \left[1 + n \int_0^\infty 4\pi r \varepsilon(r) \frac{\sin sr}{s} \, dr \right] \qquad (1)$$

with

$$s = \frac{4\pi \sin (\theta/2)}{\lambda}.$$

To vary s we can adjust θ and the wavelength λ. We put

$$\frac{I}{Nf^2} - 1 = i(s). \qquad (2)$$

We obtain

$$si(s) = n \int_0^\infty 4\pi r \varepsilon(r) \sin sr \, dr. \qquad (3)$$

This Fourier transform can be inverted to give

$$r\varepsilon(r) = \frac{1}{2\pi^2 n} \int_0^\infty si(s) \sin sr \, ds. \qquad (4)$$

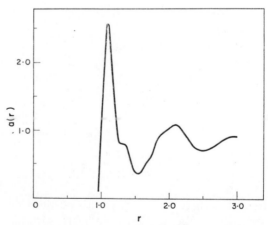

Fig. 6.3. $a(r)$ for liquid argon at 84.5°K (Dasannacharya and Rao, 1965). Measurements made at Chalk River with neutrons.

Figure 6.3 shows the results of measurements made on liquid argon. The atoms clearly show their impenetrability and their tendency to accumulate in each other's vicinity: this is the effect of the van

112

der Waals forces. All the neighbouring atoms co-operate in this accumulation which is considerably attenuated at lower densities or at higher temperatures.

Measurements at small angles—or at long wavelengths—allow us to obtain a quantity which appears in fluctuation theory:

$$i(0) = n \int_0^\infty 4\pi^2 \varepsilon(r)\, \mathrm{d}r = n \int \varepsilon(r)\, \mathrm{d}^3 r\,; \qquad (5)$$

$i(0)$ is thus connected with the isothermal compressibility:

$$i(0) = nkT\chi - 1\,.$$

There is no question of extrapolating the formalism as far as visible light because of the multiple molecular scatterings which are then superimposed to form the scattered light.

Long-wavelength neutrons permit a similar exploration. To obtain the correlations between nucleons in an atomic nucleus high-energy electrons have to be used.

6.7. Van der Waals Equation of State

To use Ursell's equation in experimental comparisons an explicit force law has to be introduced. The force laws which are operative between the molecules are not known with much accuracy and even if they were we should have some difficulty in making rigorous statements. This is why in an attempt of this kind somewhat unsophisticated considerations have to reign in a first approach to the question.

It is usual to put

$$n = N_0/V, \qquad (1)$$

N_0 being the Avogadro number and V the volume of one mole. The Ursell equation of state then takes the following form:

$$PV = RT\left(1 + \frac{B}{V} + \frac{C}{V^2} + \frac{D}{V^3} + \cdots\right) \qquad (2)$$

with

$$B = -\tfrac{1}{2}N_0 \int g_{12}\, \mathrm{d}^3 r_2, \qquad (3)$$

$$C = -\tfrac{1}{3}N_0^2 \int g_{12}g_{13}g_{23}\, \mathrm{d}^3 r_2\, \mathrm{d}^3 r_3, \qquad (4)$$

. . . .

where B is the second virial coefficient, C the third one, and so on.

113

The simplest idea is to treat the molecules as hard spheres. Their diameter is denoted by σ. The corresponding W and g are represented by Fig. 6.4:

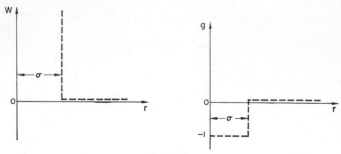

FIG. 6.4.

The virial coefficients are independent of the temperature. Calculation gives the following results (Ree and Hoover, 1967):

$$B = \frac{2\pi}{3} N_0 \sigma^3 = b_0, \qquad C = \frac{5}{8} b_0^2,$$

$$D = 0.2869 \, b_0^3,$$

$$E = (0.1103 \pm 0.0003) \, b_0^4,$$

$$F = (0.0386 \pm 0.0004) \, b_0^5,$$

$$G = (0.0138 \pm 0.0004) \, b_0^6. \tag{5}$$

We therefore have

$$PV = RT\left[1 + \frac{b_0}{V} + \frac{5}{8}\left(\frac{b_0}{V}\right)^2 + 0.2869\left(\frac{b_0}{V}\right)^3\right.$$

$$\left. + 0.1103\left(\frac{b_0}{V}\right)^4 + \cdots\right]. \tag{6}$$

When the molecules are packed so tight that they can no longer move the pressure becomes infinite. This phenomenon occurs, very roughly, when the volume is reduced to the value

$$b = b_0/3. \tag{7}$$

The series (6) should therefore diverge when the volume approaches the value b. We therefore expect as a rough approximation of (6):

$$P = \frac{RT}{V - b}. \tag{8}$$

The formula proposed by Rocard (1952),

$$P = \frac{RT}{(V - b)^3} V^2, \tag{9}$$

follows the series expansion much better. We should point out that the method which consists of replacing a limited expansion by an equivalent fraction, called the Padé approximant method, is convenient and effective (Baker, 1965).

We note that some calculations show that a segregation occurs when the given volume is close to the minimum and that relatively regular packings appear before the limit: the "solid" formed in this way naturally has no cohesion.

All in all, it appears that a gas of hard spheres is less compressible than a perfect gas. Real gases behave in this way at high temperatures but the situation is different at low temperatures. The second virial coefficient, which is at first positive, becomes zero and changes sign when the temperature falls. This phenomenon can be explained by the existence of attractive forces which are exerted beyond the repulsive forces which are responsible for the impenetrability of the molecules. Van der Waals found it convenient to assume that these attractive forces are very weak but have a relatively long range. The length of the range thus compensates for the weakness of the force and we thus succeed in predicting effects that are qualitatively reasonable. Progress in the theory of intermolecular forces will no longer allow us to keep to this scheme, but we shall keep to it nevertheless in our study which will continue to be concise.

In the following g denotes the g of the hard spheres and δg the perturbation caused by the van der Waals forces δW. To a first order we have

$$\delta g = \delta W / kT. \tag{10}$$

In these conditions B given by equation (5) is subject to a perturbation

$$\delta B = \frac{2\pi N_0}{3kT} \int_\sigma^{\sigma'} \delta W r^2 \, dr. \tag{11}$$

The integration is extended from σ to σ', this last quantity being the extreme range of the force. δW is very small, σ' is large although small when compared with the given volume, δB is a finite quantity.

115

For the third virial coefficient the first-order correction can be written as

$$\delta C = \frac{N_0^2}{kT} \int (\delta W_{12})\, g_{13} g_{23}\, \mathrm{d}^3 r_2\, \mathrm{d}^3 r_3, \tag{12}$$

the domains of integration being limited here not by the range σ' but by the condition that the unperturbed factor

$$g_{13} g_{23}$$

is non-zero, which considerably reduces the importance of δC. The same is the case for the other virial coefficients. To a first approximation we retain only the perturbation of B. We put

$$\delta B = -a/RT; \tag{13}$$

a is positive and independent of the temperature. The second virial coefficient therefore decreases at the same time as the temperature.

Let us return to the equation of state seen as a whole. For the unperturbed series we keep the unsophisticated expression in $V/(V - b)$ and we add to it the correction which results from formula (13), namely $-a/RTV$. We obtain

$$P = \frac{RT}{V - b} - \frac{a}{V^2}. \tag{14}$$

This is the famous van der Waals equation.

6.8. Second Virial Coefficient

Experimental study of the compressibility of gases at moderate densities allows us to determine the first few virial coefficients. We shall limit ourselves here to studying the one which predominates, i.e. to examining the properties of B which have already been mentioned in the last section.

It can, in general, be written as

$$B = -\frac{2\pi}{3} N_0 \int g r^2\, \mathrm{d}r. \tag{1}$$

The hypothesis of hard spheres combined with a weak long-range attraction leads, in particular, to

$$B = b_0 - \frac{a}{RT}. \tag{2}$$

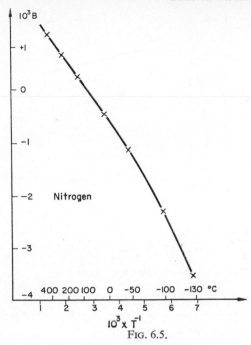

FIG. 6.5.

This last formula corresponds qualitatively to observation as shown by the figure above relating to nitrogen.† The measurements are limited at the low temperature end by the approach of condensation. A certain curvature shows up at high temperatures: the difficulty of the measurements stops the data around 400°C. The law of corresponding states predicts that the phenomenon which appears around 400°C should be observable at lower temperatures for helium: this is shown effectively by the Fig. 6.6. The virial coefficient shows a maximum at high temperatures (here around −100°C) then decreases quite slowly. This is due to the fact that the molecules are not impenetrable and that the concept of hard spheres must be replaced by the idea of repulsive forces, which are strong but vary continuously.

A satisfactory semi-empirical force law has been suggested by Lennard-Jones. It can be formulated as

$$W = 4\varepsilon \left[\left(\frac{\sigma}{r} \right)^{12} - \left(\frac{\sigma}{r} \right)^{6} \right], \tag{3}$$

† The corresponding data are due to Holborn and Otto (1926).

117

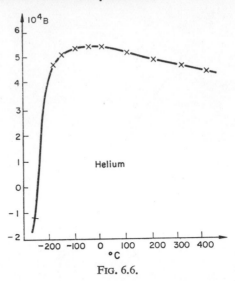

FIG. 6.6.

the parameters ε and σ having to be suitably chosen for each case. The energy here decreases much more rapidly with distance than in the case of the van der Waals hypothesis, which agrees with everything we can know about the interactions between neutral molecules. For a definite substance the available experimental data do not always cover the desirable temperature range. Fortunately theory and experiment agree here on a fundamental point: the law of corresponding states, which can be applied very well to the data relating to the second virial coefficient, is on the other hand a consequence of the theory when we assume that the force law depends only on two arbitrary parameters, as is the case for Lennard-Jones's law. A single curve in reduced coordinates shows us that formula (1) completed by (3) reproduces the facts well (Hirschfelder, Curtiss and Bird, 1964, p. 164).

It should not be forgotten that the exploration of very low temperatures cannot be carried out without introducing quantum considerations (Hirschfelder, Curtiss and Bird, 1964, Chapter 6 by J. de Boer and R. Byron Bird).

6.9. Various Calculations

Apart from the hard spheres case mentioned in section 6.7 the calculation of the virial coefficients has not been the subject

118

of very systematic work. The reasons for this are: (1) the complication of the problem, (2) the uncertainty which occurs in the choice of a good force law,† (3) an hypothesis to which we have been very faithful because of its convenience but which is not rigorous and consists of stating that the intermolecular force between two molecules is the same whether these molecules are isolated or surrounded by others of the same kind (Dymond, Rigby and Smith, 1965), (4) the fact also that the calculation of a few terms only of a series which is not particularly convergent can be slightly misleading.

This gap is nevertheless regrettable for several reasons: on the one hand the substitution of a fraction for the series in accordance with the procedures which we mentioned in section 6.7 certainly helps to improve the results of the calculation. On the other hand, the theory of the critical point—of which we shall speak later—cannot progress until these calculations have been made. As the critical densities are relatively low it would probably be enough to go as far as the fourth virial coefficient and to have some information on the fifth.

At present we must be content with overall considerations and partially qualitative deductions.‡

Other calculations relate to the double density or—equally well—to the dimensionless quantity $a(r)$. This quantity is given in principle by formula (7) of section 5.4. In any case experiment tells us that this function of the distance has a "range" which is relatively great when compared with the range of the forces. The range of the successive terms in the above expansion is equal to the range of the forces multiplied by the rank of the term in question. It is desirable to find a more convergent calculation procedure. We try to do this by combining two formulae of variations. When we make the applied field vary we have obtained on the one hand, in (7) of section 5.4, a first relation

$$\delta n_1 = -\beta n_1\, \delta V_1 - \beta \int (n_{12} - n_1 n_2)\, \delta V_2\, d^3 r_2 ; \qquad (1)$$

on the other hand, by varying the generalized Laplace equation we obtain

$$-\beta\, \delta V_1 = (\delta n_1/n_1) - \int L_{12}\, \delta n_2\, d^3 r_2 \qquad (2)$$

† The reader can be referred to recent papers (Kingston, 1965; Dymond, Rigby and Smith, 1965).

‡ It will be of interest to the reader to refer to the works by Rocard (1952) and by Hirschfelder, Curtiss and Bird (1964).

in which we have put

$$L_{12} = g_{12} + \int \tfrac{1}{2} \triangleright \, n_3 \, d^3r_3$$

$$+ \int \left(\tfrac{1}{2}\tfrac{1}{2}\varnothing + \tfrac{1}{2}\tfrac{1}{2}\boxtimes + 1 \diamondsuit^2 + \tfrac{1}{2} 1 \diamondsuit 2 + 1 \diamondsuit^2 + \tfrac{1}{2} 1 \diamondsuit 2 + 1 \diamondsuit^2 \right)$$

$$\times \, n_3 n_4 \, d^3r_3 \, d^3r_4 + \cdots . \tag{3}$$

The expansion that we have just written down does not include between the points 1 and 2 the open chains of the expansion (10) of section 6.1. It is tempting to think that it is better from the point of view of range and from the point of view of convergence.

This stated, let us note that equations (1) and (2) represent two linear transformations that are the inverse of each other, the first allowing us to change from δV to δn, and the second from δn to δV. There are clearly relations between the elements of these two transformations.

Let us use the notation which allows us to condense the relative formulae to linear operations. The formulae (1) and (2) can be written symbolically as

$$\delta n/n = \mathscr{M}(-\beta \, \delta V), \tag{4}$$

$$-\beta \, \delta V = \mathscr{L}(\delta n/n), \tag{5}$$

$\mathscr{M}$ and $\mathscr{L}$ being two linear operators which are the inverse of each other:

$$\mathscr{L}\mathscr{M} = 1. \tag{6}$$

They are real Hermitian operators. $\mathscr{M}$ is a bounded operator: its eigenvalues are discrete. They are positive. Let us consider an arbitrary, possibly even complex, function of a point

$$F(x_J) \quad \text{or} \quad F_J.$$

The average of $\sum_J F_J$ is

$$\int n_1 F_1 \, d^3r_1 .$$

The variance of the same quantity can be written explicitly as follows:

$$\langle |\varDelta \sum_J F_J|^2 \rangle$$

$$= \int n_1 F_1^* F_1 \, d^3r_1 + \int (n_{12} - n_1 n_2) \, F_1^* F_2 \, d^3r_1 \, d^3r_2 . \tag{7}$$

It is positive—exceptionally zero—which justifies the assumption we just made about the eigenvalues.

The operator $\mathscr{L}$ has the same eigenfunctions as $\mathscr{M}$. Its eigenvalues are the reciprocals of the eigenvalues of $\mathscr{M}$.

In this chapter we are interested in uniform media. It is convenient to assume that they extend to infinity. The set of the eigenfunctions and of the eigenvalues becomes continuous. The eigenfunctions are plane waves

$$(2\pi)^{-3/2} \exp ik_x x,$$

which are normalized here. The eigenvalue of $\mathscr{M}$ which corresponds to the above wave is

$$\mathscr{M}_k = 1 + 4\pi n \int_0^\infty \varepsilon(r) \frac{\sin kr}{k} r \, dr. \tag{8}$$

The modulus of the vector k has been denoted by k. We have returned to the notation

$$n_{12} - n_1 n_2 = n_1 n_2 \varepsilon_{12}.$$

Here ε depends only on the distance. Likewise the eigenvalues of $\mathscr{L}$ are given by the formula

$$\mathscr{L}_k = \mathscr{M}_k^{-1} = 1 - 4\pi n \int_0^\infty L(r) \frac{\sin kr}{k} r \, dr. \tag{9}$$

The equation

$$\mathscr{L}_k = 0 \tag{10}$$

generally has no root, since, except in exceptional cases, the correlations decrease so rapidly that $\mathscr{M}_k$ has only finite values.

In order to profit from the general theorems we put the transformation (1) in the following form:

$$\delta n_1/n = \int [\delta(r_2 - r_1) + n\varepsilon(r)] (-\beta \, \delta V_2) \, d^3 r_2 \tag{11}$$

where the three-dimensional Dirac function appears. The relation which follows results then from the properties of the linear operators:

$$\delta(r_2 - r_1) + n\varepsilon(r) = (2\pi)^{-3} \int \exp ik_x(x_2 - x_1) \mathscr{M}_k \, d^3 k, \tag{12}$$

or

$$\varepsilon(r) = (8\pi^3 n)^{-1} \int_0^\infty 4\pi \frac{\sin kr}{r} (\mathscr{M}_k - 1) k \, dk, \tag{13}$$

or, introducing $\mathscr{L}_k$,

$$\varepsilon(r) = (2\pi^2 nr)^{-1} \int_0^\infty \sin kr \frac{1 - \mathscr{L}_k}{\mathscr{L}_k} k \, \mathrm{d}k. \tag{14}$$

This formula is particularly useful for studying the behaviour of the correlations at great distances. By virtue of the Fourier transform the behaviour of ε is controlled by the poles of $\mathscr{L}_k$, k now being treated as a complex variable which lies in the upper complex half-plane. The asymptotic properties in particular are controlled by what happens for small values of k. To pick out the poles closest to $k = 0$ it is legitimate to expand $\mathscr{L}$ into a series and to write

$$\mathscr{L}_k = \mathscr{L}_0 + k^2 \mathscr{L}_{\mathrm{II}} + \cdots \tag{15}$$

with

$$\mathscr{L}_0 = 1 - 4\pi n \int_0^\infty L(r) \, r^2 \, \mathrm{d}r, \tag{16}$$

$$\mathscr{L}_{\mathrm{II}} = \tfrac{2}{3} \pi n \int_0^\infty L(r) \, r^4 \, \mathrm{d}r. \tag{17}$$

By going back to the definition of L and to the equation of state we can verify the following relation:

$$\mathscr{L}_0 = \beta(\partial P/\partial n) = (nkT\chi)^{-1}, \tag{18}$$

χ denoting the compressibility. Equation (10) therefore has a real root, and a zero root when the compressibility coefficient is infinite. This is what occurs when the fluid is in the critical state.

In the vicinity of the critical point the expansion (15), limited to its first two terms, is sufficient to supply the smallest root of (10). It is necessarily a complex root. We shall put

$$\mathscr{L}_{\mathrm{II}} = \Lambda^2, \tag{19}$$

Λ being a certain length characteristic of the fluid under consideration. In the vicinity of the critical point the asymptotic behaviour of the correlations is given by the following formula:

$$\varepsilon = (4\pi^2 inr)^{-1} \int_{+\infty}^{-\infty} \frac{1}{\mathscr{L}_0 + \Lambda^2 k^2} k \exp(ikr) \, \mathrm{d}k, \tag{20}$$

122

or

$$\varepsilon = \frac{1}{4\pi n \varLambda^2} \frac{\exp(-k_0 r)}{r}, \text{ with } k_0^{-1} = (nkT\chi)^{1/2} \varLambda. \quad (21)$$

Because of continuity we can extend this method of reasoning further from the critical point, provided that a larger number of terms is retained in the expansion (15). In the next approximation equation (10) becomes

$$\mathscr{L}_{IV} k^4 + \mathscr{L}_{II} k^2 + \mathscr{L}_0 = 0, \quad (22)$$

hence

$$k^2 = [-\mathscr{L}_{II} \pm (\mathscr{L}_{II}^2 - 4\mathscr{L}_0 \mathscr{L}_{IV})^{1/2}]/2\mathscr{L}_{IV}. \quad (23)$$

$\mathscr{L}_0$ is positive. It can be assumed that $\mathscr{L}_{II}$ has stayed so because of its continuity. As this equation can have no real root, $\mathscr{L}_{IV}$ is also positive. Let us therefore rewrite (23) taking these remarks into consideration:

$$k^2 = -(\varLambda^2/2\mathscr{L}_{IV})\{1 \pm [1 - 4(\mathscr{L}_0 \mathscr{L}_{IV}/\varLambda_4)]^{1/2}\}.$$

Finally, the only quantity to have rapid variations in the vicinity of the critical point, as a function of the density and of the temperature, is probably $\mathscr{L}_0$. We therefore assume that only this quantity has a singularity at the critical point. On the gas side $\mathscr{L}_0$ remains small enough for the roots to remain purely imaginary. The asymptotic law remains similar to equation (21). On the other hand, since the compressibility decreases, it may happen that the quantity under the radical becomes negative: the asymptotic decay law will therefore be of the damped sine-wave type.

These remarks, which are based on very unsophisticated arguments, agree qualitatively with observation. We have already mentioned the tendency of the molecules of a liquid each to be surrounded with concentric layers of other molecules, this association becoming more and more vague when the distance increases. This is the start of a crystal lattice. This is a phenomenon which is accentuated when the temperature falls and which is not limited in any way, as is the case of the arguments presented above, to the vicinity of the critical point.

These phenomena are due to the van der Waals forces. The above considerations are half-theoretical and half-empirical. Their weak point is the hypothesis that $L(r)$ decreases rather rapidly with the distance to justify the expansion (15).

Fine Structure in Thermal Equilibrium

The subject touched on in this section is at present the subject of very active work (Frisch and Lebowitz, 1964; Verlet and Levesque, 1967). From this we can expect, after a certain amount of adjustment, the method of approximation which was missing some years ago and which the theory of relatively dense fluids needs.

CHAPTER 7

Phase Changes

7.1. Condensation

The phenomena of compressibility and condensation can be followed from the network of isotherms where we can distinguish (see Fig. 7.1):

(1) high temperatures with isotherms of a perfect or almost perfect gas;

(2) the critical isotherm with its horizontal inflexion point

$$\partial P/\partial V = 0 \quad \text{or} \quad \chi = +\infty;$$

(3) the liquefaction curve;

(4) an isotherm with its liquefaction horizontal; on it $\partial P/\partial V = 0$ but there are two phases;

(5) the extension of this isotherm towards the metastable states of a superheated liquid and a supersaturated vapour.

Recent experiments (Saltsburg, 1965) have led to the observation of superheating of cadmium. The normal boiling point of the metal is $1040°K$. Superheating of the order of $1000°$ has been achieved. On the other hand, the metastable liquid can reach negative pressures. Experiments with mercury have been able to go as far as -425 atm.

The phenomena are described qualitatively by the van der Waals equation, provided we know how to put in the liquefaction horizontal. Nevertheless accuracy in details cannot be asked from a theory which describes the forces as approximately as this one does.

In a uniform medium, except at the critical point, the isothermal compressibility coefficient is always positive. This corresponds to our formula which connects this coefficient with the fluctuations

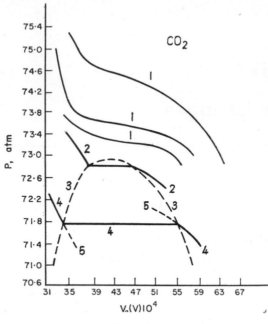

FIG. 7.1.

and which says it must be positive:

$$\Delta N^2/N^2 = (1/N)\,nkT\chi. \tag{1}$$

The van der Waals equation includes both unstable states and stable states. This possibility originates from the non-linearity of the Laplace equation generalized with respect to the density. This equation cannot be used by itself to distinguish between stable and unstable states. The condition

$$\chi > 0$$

allows us to select the stable and metastable states. To distinguish the stable states from the metastable states we must add to it another stability criterion. We shall leave this problem on one side with the reminder that classical thermodynamics settles this problem completely and that it would be sufficient to bring in its arguments with a seasoning of statistical considerations.

This said, to develop the theory of condensation it is convenient to imagine a very large isothermal container of great height acted

126

upon by the field of gravity

$$Z_1 = -mg. \tag{2}$$

We can thus find the extremes of pressure and of density on the same vertical. Figure 7.2 is drawn for the case where there is condensation and the liquid does not wet the wall.

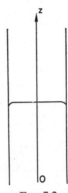

FIG. 7.2.

It is unwise to do away with gravity although at first sight it seems to play an unimportant part. Its absence would be risking the appearance of neutral equilibria. Theory in its conscientious way would give us the average effect of these equilibria: an effect unrelated to observation. If we eliminate all force fields we prevent the fluid from having a definite density at each point [cf. formula (3) of section 5.4]. Although space research is expected to give us the opportunity of observing phenomena of this kind it is preferable to deal with a more familiar situation.

Let us therefore return to our fluid with weight. We follow a vertical far enough from the wall for the interphase which it eventually meets to be horizontal.

All along this vertical symmetry requires that the off-diagonal components of the pressure are zero:

$$P_{xy} = P_{xz} = P_{yz} = 0, \tag{3}$$

and that the components P_{xx} and P_{yy} are equal. As for the density, it is connected with the last component by the following relation:

$$-mgn = \frac{\partial P_{zz}}{\partial z}. \tag{4}$$

Above the interface the density decreases fairly slowly from the density of the saturated vapour but becomes zero at great height. Below the interface it increases even more slowly from the density of the surface liquid but becomes extreme at great depths.

In the interface, over a distance of a few Ångströms, the density passes through all intermediate values. The spatial variation here is extremely rapid.

Outside the interface the fluid is practically uniform:

$$P_{xx} = P_{yy} = P_{zz} = P. \tag{5}$$

In the interface it is definitely not. This equality no longer exists. As we know that surface tension tends to reduce the surface of the interface it must even be assumed that the inequality

$$P_{xx} < 0 \tag{6}$$

predominates almost everywhere. The surface tension is the integral of this component through the interface with the opposite sign:

$$A = - \int_{z_1}^{z_2} P_{xx} \, dz. \tag{7}$$

The sign of the surface tension means that the force law includes attractive forces which predominate in the calculation of A. In the absence of attractive forces there could be no condensation. Let us recall, moreover, that it is they that introduce the term $-a/V^2$ into the van der Waals equation. If the sign of this term were changed the isotherm network would no longer have the desired properties.

In the interface by virtue of equation (4) P_{zz} hardly varies at all. It is the pressure which obtains in the gas, or the imperceptibly higher pressure which obtains in the liquid in the immediate vicinity of the interface.

In the interface P_{xx} is very much higher than this pressure. For example, for mercury at 0°C *in vacuo*

$$A = 480 \text{ dyne cm}^{-1}.$$

Let us assume for the sake of simplicity that this quantity corresponds to a pressure which would be exerted uniformly in the interface at a height of 5 Å, let us say, then the corresponding value of P_{xx} is

$$P_{xx} = -9600 \text{ atm},$$

i.e. very much greater than we have been able to observe with metastable mercury.

Surface tension poses a very interesting problem on which we shall touch only briefly here and the approach to which should be supplied by the generalized Laplace equation.

In any case this equation allows us to determine the densities on each side of the interface, the vapour side and the liquid side. In the same way as P_{zz}, the two terms of the generalized Laplace equation hardly vary at all across the interface. Hence for the two uniform media in question the two conditions required are

$$P_V = P_L, \tag{8}$$

$$\Phi_V = \Phi_L. \tag{9}$$

The second of these conditions simply expresses the fact that the two phases have the same chemical potential. Φ is the second term of the Laplace equation in which we assume that the density is constant.

It is true that our study omits the explanation of the fact that the interface is only very thin. This is a problem to be treated at the same time as that of the surface tension. It is quite obvious, on the other hand, that our interest is limited to a static view of the phenomena: the dynamics of condensation and vaporization are based on more ambitious considerations.

7.2. Critical Point

The critical point corresponds to the top of the condensation curve or, equally, to the only isotherm without a horizontal section, but with one point with a horizontal tangent.

There the compressibility coefficient is infinite, whilst the medium is uniform. In accordance with our formula for the fluctuations of densities—or of the number of particles present in a given volume—the fluctuations become infinite:

$$\frac{\langle \Delta N^2 \rangle}{\langle N \rangle^2} = \frac{1}{k} kT\chi. \tag{1}$$

We can see that the fluctuations cannot really be infinite but that nevertheless they become very large. Observation in white light shows that the medium scatters the light strongly: this scattered light has the appearance of the blue of the sky and is very characteristic. This phenomenon must not be confused with the milky appearance which the fluid takes on when it is brought to just below the critical point: there the liquid and the vapour have almost the same density and the droplets of liquid, although helped by gravity, collect together only slowly.

Microscopic analysis of the fluctuations led us to the formula

$$\langle \Delta N_A^2 \rangle = \int n_1 \, d^3r_1 + \int_A \int_A (n_{12} - n_1 n_2) \, d^3r_1 \, d^3r_2,$$

or, in a uniform medium neglecting a surface effect,

$$\langle \Delta N_A \rangle = V_A \left(n + 4\pi \int n^2 \varepsilon(r) \, r^2 \, dr \right), \tag{2}$$

where the fluctuations can become very great only if $\varepsilon(r)$, contrary to what happens in the present circumstances, becomes zero only at very large distances. By using an, in fact, only moderately well justified hypothesis we were able to show in section 6.9 that in the vicinity of the critical point, in a uniform phase,

$$\varepsilon_{12} \sim (e^{-kr_{12}}/r_{12}),$$

k, ideally, becoming zero at the critical point. But a formula like

$$\varepsilon_{12} \sim 1/r_{12}$$

has no meaning except for an infinite medium, which will never correspond to the experimental conditions.

Let us take up our observations in a gravity field again. Before the critical temperature is even reached gravity flattens out the fluid in the region where it is very compressible and favours the appearance of the meniscus at higher temperatures than the ideal critical temperature. A horizontal and uniform layer where χ would be effectively infinite cannot exist; it is flattened out by the layers just above it. The density fluctuations between two horizontal layers can be calculated by formula (8) of section 5.6 and it can be shown that they always remain finite.

Observation of the phenomena very close to the critical point presents considerable difficulty since the least error in the temperature or the least impurities falsify things completely. Certain physicists have observed hysteresis phenomena; this is comprehensible if the fluid wavers there between the stable state and metastable states. But it is legitimate to query the description of what one believes to have seen. Equilibrium is established only slowly. Measurements of density and pressure must allow for the non-uniformity due to gravity. Extrapolations in the direction of the ideal critical point are delicate.

The situations where the correlations tend—exceptionally—to extend a long way are very interesting. A critical point is observed at the maximum temperature where liquid mixtures, which mix in any proportion only at high temperatures (for example), separate. The mixture of methyl alcohol and carbon sulphide can be quoted as an example.

The same phenomenon can also be observed in solid solutions: it has been shown in an aluminium–zinc alloy (Münster, 1959) (atomic percentage of zinc $40^0/_0$, critical temperature 351.5°). In these non-transparent media observations are made with X-rays. The measurements must be made at small angles.

In ferromagnetic substances the correlations no longer relate to the positions of the atoms (or the ions) which, if we neglect thermal motion, are fixed at the lattice sites, but to the relative orientations of the atomic magnetic moments. At the ferromagnetic critical point, i.e. at the Curie point, major fluctuations appear (Van Hove, 1954a and 1954b). They are observed by neutron scattering.

The van der Waals equation, or similar more elaborate equations, describe the critical isotherm as an isotherm which has a horizontal tangent at its inflexion point. They ascribe no other particular property to it. We can deduce from it because of continuity that in the vicinity of the critical temperature the density of the liquid and the density of the vapour in equilibrium are connected by a relation of the following type:

$$n_L - n_V \sim [1 - (T/T_c)]^{1/2}.$$

By a systematic application of the principle of corresponding states Guggenheim (1945) obtained from the experimental data a different law which substitutes the exponent $\frac{1}{3}$ for the exponent $\frac{1}{2}$. The critical point would then have more singular properties than have been formerly attributed to it.

This possibility is reinforced by a whole series of experimental results which relate to the actual shape of the critical isotherm, the isothermal compressibility coefficient above the critical temperature and the specific heat at constant volume (Fisher, 1964).

A theoretical interpretation has been suggested for these phenomena (Fisher, 1964): it is based on a similarity argument and on

131

what can be estimated from the properties of a model which is more convenient to deal with than the real problem; in this model the molecules do not move freely in space but can only jump from one site to another on a regular lattice.

This model is a little artificial. It can probably be improved. If we keep, however, to the orthodox description where the molecules have no privileged positions the crux of the question seems to be in the study of the behaviour at great distances of the quantity $L(r)$ which we defined in section 6.9 and which we treated in the "traditional" way in that section.

We come back to the interpretation of the critical point: except perhaps for the immediate vicinity of the critical point the theory of condensation, based on the equality of the pressures and the chemical potentials, is not modified in any way.

7.3. Surface Tension

This section is limited to some basic considerations.

The theory of condensation is unavoidably connected with the theory of surface tension. We have assumed rather hastily that the generalized Laplace equation would correctly supply the description of the interface and that it would explain in particular the very rapid variation in density between the two phases.

We return to the enormous container with vertical walls where the liquid rests beneath its vapour. The interface is practically horizontal. Gravity intervenes only to bring order into the phenomenon: the liquid and the vapour are uniform. P_{zz} is constant throughout the container and P_{xx} is a function of the height z which becomes the same as P_{zz} outside the interface.

In this case the relation between the grand partition function and the pressure which is suitable for a uniform medium can be generalized to the following form

$$\ln \Xi = \beta \left[V P_{zz} + S \int (P_{xx} - P_{zz}) \, \mathrm{d}z \right]. \tag{1}$$

Here V is the volume of the container and S is its cross-section. The integral

$$A = \int (P_{zz} - P_{xx}) \, \mathrm{d}z, \tag{2}$$

whose limits it is useless to specify, is in fact limited to the interface. In situations when the surface tension is strong formula (2) is equivalent to formula (7) of section 7.1. In any case it is more satisfactory and it must be considered as the true definition of the surface tension.

The problem of surface tension is daunting since it consists of solving the Laplace equation—by neglecting the applied field, then of calculating the integral (2), for example via (1).

But there is a variational principle which can be used. Let us return to the general case of a system in thermal equilibrium in some force field. The quantity

$$\ln \varXi + \int n_1 \left[\alpha + \ln \frac{(2\pi mkT)^{3/2}}{\Gamma} - \beta V_1 - \varPhi_1 \right] d^3r_1, \qquad (3)$$

where $\varPhi_1$ represents the second term of the Laplace equation, is the same as $\ln \varXi$ when the Laplace equation is satisfied.

We consider $\ln \varXi$ as a functional of the density and of the inter-molecular energy. Let us leave the latter unchanged. From (3) of section 6.3 it follows that

$$\delta \ln \varXi = \int n_1 \, \delta \varPhi_1 \, d^3r_1. \qquad (4)$$

As a consequence, if we keep the temperature, the chemical potential, the applied energy and the intermolecular energy constant, the quantity (3) is extremal with respect to the density when the Laplace equation is satisfied. An additional result of the properties of L_{12} is that the second variation is necessarily negative (de Dominicis, 1962).

These properties show up some favourable prospects for the solution of the Laplace equation in a non-uniform medium and of the calculation of the surface tension. There is only a first indication there and a better choice is perhaps possible in the gamut of extremes that can be imagined (de Dominicis, 1962).

These extremal properties help us to understand why points of view as contradictory as those of Rocard (1952)—the variation of the density is treated as an analytical function of the height—and of Kirkwood and Buff (1949)—the variation is treated as discontinuous—both lead to numerically acceptable results.

CHAPTER 8

Thermal Waves in Fluids

8.1. Introduction

A fluid in thermal equilibrium such as we have described up to the present is a medium where order obtains. It closely obeys the laws of thermodynamics. The motion of the molecules is carefully camouflaged. The point is that our study has been carried out to meet the needs of moderately refined experiments.

Thermal motion is brought out particularly by the observation of the movement of mycelia put in suspension in the fluid (Brownian motion) and also by the observation of the diffusion of light. We should mention here, *inter alia*, phenomena whose observation alters the state of the system studied as little as possible.

Nevertheless the study of Brownian motion assumes the introduction of foreign bodies. The effects of light—the medium is assumed to be transparent—are imponderable. It is nevertheless necessary that the intensity should be fairly weak so as to avoid the occurrence of mechanical effects and of non-linear phenomena. When a bundle of light rays passes through a transparent medium each molecule scatters a fraction of the incident light; the larger part of the scattered light combines with the incident light to form the coherent wave. The rest originates in the coherent wave and escapes in all directions. This is scattered light.

The scattered radiation consists of Rayleigh radiation which—assuming the incident light to be monochromatic—has the same wavelength as the incident radiation, within a line width or so, and of Raman radiation which corresponds to wavelengths clearly different from those of the source. In monatomic media, which are those we are considering here, the Raman scattering is insignificant.

Brillouin (1922) studied the interaction between an elastic wave and an electromagnetic wave. He showed that light is scattered by elastic waves as by a lattice. The scattered ray has a doublet structure. The structure of the Rayleigh ray—whose precise observation is still difficult even at the present time—shows a similar structure, which leads to thermal motion being represented in the form of wave trains superimposed on each other. We intend to devote the present chapter to the molecular theory of these motions.

We are dealing only with fluids here. The case of crystal lattices was illustrated long ago by Einstein and by Debye. In the lattices elastic waves (and all kinds of waves which affect electrons and magnetism) are clearly a means of describing the internal motions. In liquids this concept is more artificial. The nature of the subject will oblige us to use specific methods which may later be useful in problems that appear to be more accessible.

8.2. Density Waves

In this chapter we do not consider condensation phenomena. It is useless to bring in an applied field. The fluid is uniform and isotropic. We assume that the fluid is contained in a cube of volume V. We shall expand the quantities to be considered with the help of complex elementary waves which are very convenient to use.

$$\varphi_s(r) = V^{-1/2} \exp\left[i\,(s \cdot r)\right], \tag{1}$$

where s is the wave vector, r the vector which marks the position of a point in the cube.

The walls of the cube are no longer walls in the usual sense. The fluid is not properly enclosed in the cube, it is subject to cyclic conditions. Each time a particle coming from outside penetrates a face of the cube at a certain velocity another molecule leaves the cube at the same velocity through the opposite face at the corresponding point. The number of particles contained in the cube is thus constant. Nothing prevents the fluid from being exactly uniform. The cyclic conditions replace the closure conditions that we adapted in section 1.8. They allow the same integrations by parts. The wave vectors form a discrete set determined by the condition that the waves must be periodic between the faces of the cube. The cube is also large

135

enough for the wave vectors to form a practically continuous set in the range of wavelengths of interest to us. These vectors can be in any direction, but every time that it is possible we shall assume the vector to be along Ox, this coordinate axis being parallel to an edge of the cube. Since the grain of the fluid has a much finer structure than the dimensions of the cube this direction has no special physical property.

The waves (1) form a complete orthonormal system.

Let us return to the microscopic density introduced by equation (1) of section 1.5:

$$\nu(r) = \sum_1^N \delta\,(r - r_J). \tag{2}$$

We expand it in a Fourier series:

$$\nu(r) = \sum_s B_s \varphi_s(r), \tag{3}$$

from which we obtain a density amplitude

$$B_s = V^{-1/2} \sum_J \exp\,[-\,\mathrm{i}(s \cdot r_J)]. \tag{4}$$

It is clear that

$$B_{-s} = B_s^*.$$

The average value of B_s calculated in the framework of the canonical distribution and for a uniform medium is obviously zero. To find interesting results the square of the density amplitude must be calculated:

$$|B_s|^2 = V^{-1} \sum_J \exp\,[-\,\mathrm{i}(s \cdot r_J)] \sum_K \exp \mathrm{i}(s \cdot r_K). \tag{5}$$

We shall have many expressions of this kind to expand. We must therefore proceed systematically. The symbol

$$\sum_J \sum_K$$

represents a sum which extends over all the possible combinations of suffices. It includes in particular the term $J = K$. On the other hand, the symbol

$$\sum_{J,K}$$

represents a sum where the case $J = K$ is excluded. It should be noted that we always have

$$\sum_J \sum_K = \sum_K \sum_J, \qquad \sum_{J,K} = \sum_{K,J}.$$

Now, before taking its average expression (5) must be transformed somewhat:

$$\sum_J \exp\left[-i(s \cdot r_J)\right] \sum_K \exp i(s \cdot r_K)$$

$$= \sum_J \sum_K \exp\left[i(s \cdot r_K - r_J)\right]$$

$$= \sum_J \sum_K \cos\left(s \cdot r_K - r_J\right)$$

$$= \sum_J \cos\left(s \cdot r_J - r_J\right) + \sum_{J,K} \cos\left(s \cdot r_K - r_J\right)$$

$$= N + \sum_{J,K} \cos\left(s \cdot r_K - r_J\right). \tag{6}$$

Let us calculate the average of this quantity in the canonical distribution. The result is as follows:

$$N + \int n_{12} \cos\left(s \cdot r_2 - r_1\right) d^3r_1\, d^3r_2, \tag{7}$$

which can be written in a more homogeneous manner, the simple density being uniform, as

$$\int n\, d^3r_1 + \int n_{12} \cos\left(s \cdot r_2 - r_1\right) d^3r_1\, d^3r_2. \tag{8}$$

It is of some advantage to change to the grand canonical distribution. We do not know how many particles the cube contains: the cyclic conditions then apply individually to each of the possible cases of occupation. The integrations by parts continue to obey the same rules. On the other hand, expression (8) remains valid. Bearing in mind the cyclic conditions and that n_{12} depends only on the distance r_{12}, this expression can also be written as

$$V\left[n + \int n_{12} \cos\left(s \cdot r_2 - r_1\right) d^3r_2\right]. \tag{9}$$

It is advantageous to introduce explicitly the difference

$$n_{12} - n^2,$$

because there is no indication that the molecular disorder hypothesis should not be applied to a uniform medium. The cyclic conditions also allow us to replace (9) by

$$V\left[n + \int \left(n_{12} - n^2\right) \cos\left(s \cdot r_2 - r_1\right) d^3r_2\right]. \tag{10}$$

Integration is now limited to a small domain. We finally obtain

$$\langle |B_s^2| \rangle = n + \int \left(n_{12} - n^2\right) \cos\left(s \cdot r_2 - r_1\right) d^3r_2. \tag{11}$$

137

Thermal Waves in Fluids

An interesting case is that of long wavelengths (L.W.)—although they are still short compared with the dimensions of the cube. We then have

$$\text{L.W.} \quad \langle |B_s^2| \rangle = n + \int (n_{12} - n^2)\, \mathrm{d}^3 r_2. \tag{12}$$

The second term is an expression that we have met many times and whose value we shall express later in terms of thermodynamic quantities, i.e. in terms of experimentally accessible quantities.

Formulae (11) and (12) can be looked upon as expressing correlations between the wave s and the wave $-s$. There are no correlations between two waves whose wave vectors do not satisfy the relation

$$s_1 + s_2 = 0. \tag{13}$$

This result can be generalized to any number of waves. Let us first consider three of them. There are no correlations unless

$$s_1 + s_2 + s_3 = 0. \tag{14}$$

In this connexion we should point out the following result which is valid in the L.W. case:

$$V^{1/2} \langle B_{s_1} B_{s_2} B_{s_3} \rangle = n + 3 \int (n_{12} - n^2)\, \mathrm{d}^3 r_2$$
$$+ \int (n_{123} - n_{12}n - n_{13}n - n_{23}n + 2n^3)\, \mathrm{d}^3 r_2\, \mathrm{d}^3 r_3. \tag{15}$$

This result will also be reduced to measurable quantities. The result does not alter if in the last integrand we replace

$$n_{12} + n_{13} + n_{23} \quad \text{by} \quad 3n_{12}.$$

In the same spirit we note further the existence of correlations between four waves the sum of whose vectors is zero:

$$\text{L. W.} \quad V \langle B_{s_1} B_{s_2} B_{s_3} B_{s_4} \rangle = n + 7 \int (n_{12} - n^2)\, \mathrm{d}^3 r_2$$
$$+ 6 \int (n_{123} - 3n_{12}n + 2n^2)\, \mathrm{d}^3 r_2\, \mathrm{d}^3 r_3$$
$$+ \int (n_{1234} - 4n_{123}n - 3n_{12}n_{34} + 12n_{12}n^2 - 6n^4)\, \mathrm{d}^3 r_2\, \mathrm{d}^3 r_3\, \mathrm{d}^3 r_4. \tag{16}$$

The recurrence equations (12), (13), (14) of section 5.3 allow the above results to be condensed. For equations (12), (15) and (16)

138

we can substitute the following results:

$$\langle |B_s|^2 \rangle = \partial n/\partial \alpha,$$
$$V^{1/2}\langle B_{s_1} B_{s_2} B_{s_3} \rangle = \partial^2 n/\partial \alpha^2,$$
$$V\langle B_{s_1} B_{s_2} B_{s_3} B_{s_4} \rangle = \partial^3 n/\partial \alpha^3.$$

8.3. Momentum

Let us take the derivation of the density amplitude with respect to time. We obtain

$$V^{1/2} (dB_s/dt) = - i(s \cdot \sum_J c_J \exp [- i(s \cdot r_J)]). \tag{1}$$

The average of this quantity is zero for two reasons: firstly because the velocity appears as a factor and secondly because of the cyclic conditions. The velocity factor has the same effect on the following average:

$$\langle B_s^*(dB_s/dt) \rangle = 0. \tag{2}$$

Taking the derivative we obtain from it the relation

$$\langle (dB_s^*/dt)(dB_s/dt) \rangle = \langle |dB_s/dt|^2 \rangle = - \langle B_s^*(d^2 B_s/dt^2) \rangle. \tag{3}$$

We can likewise establish the relations

$$\langle B_s^*(d^3 B_s/dt^3) \rangle = 0, \tag{4}$$
$$\langle |d^2 B_s/dt^2|^2 \rangle = - \langle B_s^*(d^4 B_s/dt^4) \rangle \dots. \tag{5}$$

Bearing in mind the independence of the velocities the average (3) is simply

$$\frac{1}{V} s^2 \langle \sum u_J^2 \rangle. \tag{6}$$

We therefore obtain

$$\langle |dB_s/dt|^2 \rangle = s^2 nkT/m. \tag{7}$$

In equation (1) the velocity vector is parallel to the wave vector. Waves can also be imagined where the velocity or, if one prefers it, the momentum is at right angles to the direction of propagation, which we now take along the x-axis. We shall define, for example, an amplitude

$$M_{sy} = V^{-1/2} \sum_J mv_J \exp [- isx_J] \tag{8}$$

whose mean square will be

$$mnkT. \tag{9}$$

Thermal Waves in Fluids

The momentum waves have the same mean square whether they are longitudinal or transverse.

Let us again calculate the average (3). First of all, by differentiating (1) we obtain†

$$V^{1/2} (d^2 B_s/dt^2) = - i\left(s \sum_J \left[- isc_J^2 + \frac{1}{m}\sum_L X_{JL}\right]\right) \exp [- isx_J].$$

(10)

We can likewise write

$$V^{1/2} (d^2 B_s^*/dt^2) = i\left(s \sum_K \left[isc_K^2 + \frac{1}{m}\sum_M X_{KM}\right]\right) \exp (isx_K).$$

(11)

By multiplying we obtain

$$Vs^{-2}|d^2 B_s/dt^2|^2 = \sum_J \sum_K \left[s^2 c_J^2 c_K^2 - i\left((s/m)\cdot \sum_L c_J^2 X_{KM}\right)\right.$$
$$\left.+ i\left((s/m)\cdot \sum_L c_K^2 X_{JL}\right) + m^{-2} \sum_L \sum_M X_{JL}X_{KM}\right] \exp [is(x_K - x_J)].$$

(12)

Let us first eliminate the average of the velocities. We can write

$$\sum_J \sum_K c_J^2 c_K^2 \exp [is(x_K - x_J)] = \sum_J c_J^4 + \sum_{J,K} c_J^2 c_K^2 \exp [is(x_K - x_J)].$$

The average of u^4 in the Maxwell distribution is equal to $3(kT/m)^2$. We get

$$V(m/s)^2 \langle |dB_s/dt^2|^2 \rangle_v$$
$$= 3N(skT)^2 + (skT)^2 \sum_{J,K} \exp [is(x_K - x_J)]$$
$$+ \sum_J \sum_K \left[i\left((skT \cdot \sum_L (X_{JL} - X_{KL})\right) + \sum_L \sum_M X_{JL}X_{KM}\right] \exp [is(x_K - x_J)].$$

(13)

In the last sum we group the terms by the number of suffices. It then becomes

$$\sum_{J,K} X_{JK}^2 \{1 - \exp [is(x_K - x_J)]\}\{ + \sum_{J,K,L} X_{JK}X_{JL}$$
$$+ 2i(skT \cdot \sum_{J,K} X_{JK}) \exp [is(x_K - x_J)].$$
$$+ i\left(skT \cdot \sum_{J,K,L} (X_{JL} - X_{KL})\right) \exp [is(x_K - x_J)]$$
$$+ \sum_{J,K,L} (X_{JL}X_{KL} + X_{JK}X_{KL} - X_{JL}X_{JK}) \exp [is(x_K - x_J)]$$
$$+ \sum_{J,K,L,M} X_{JL}X_{KM} \exp [is(x_K - x_J)].$$

† The formalism is satisfactory only if we make $X_{JJ} = 0$ by definition.

Allowing for the symmetry this can also be written as

$$\sum_{J,K} X_{JK}^2[1 - \cos\{s(x_K - x_J)\}] + \sum_{J,K,L} X_{JK}X_{JL}$$
$$- 2(skT \cdot \sum_{J,K} X_{JK}) \sin\{s(x_K - x_J)\} - 2(skT \cdot \sum_{J,K,L} X_{JL}) \sin\{s(x_K - x_J)\}$$
$$+ \sum_{J,K,L} (X_{JL}X_{KL} + X_{JK}X_{KL} + X_{KJ}X_{JL}) \cos\{s(x_K - x_J)\}$$
$$+ \sum_{J,K,L,M} X_{JL}X_{KM} \cos\{s(x_K - x_J)\}.$$

We then have

$$(m/s)^2\langle|\mathrm{d}^2B_s/\mathrm{d}t^2|^2\rangle = (skT)^2 \left[3n + \int n_{12}\cos\{s(x_2 - x_1)\}\,\mathrm{d}^3r_2\right]$$
$$- 2\left((skT \cdot \left[\int n_{12}\,X_{12}\sin\{s(x_2 - x_1)\}\,\mathrm{d}^3r_2 \right.\right.$$
$$\left.\left. + \int n_{123}X_{13}\sin\{s(x_2 - x_1)\}\,\mathrm{d}^3r_2\,\mathrm{d}^3r_3\right]\right)$$
$$+ \int n_{12}X_{12}^2[1 - \cos\{s(x_2 - x_1)\}]\,\mathrm{d}^3r_2 + \int n_{123}X_{12}X_{13}\,\mathrm{d}^3r_2\,\mathrm{d}^3r_3$$
$$+ \int n_{123}(X_{13}X_{23} + X_{12}X_{23} + X_{21}X_{13})\cos\{s(x_2 - x_1)\}\,\mathrm{d}^3r_2\,\mathrm{d}^3r_3$$
$$+ \int n_{1234}X_{13}X_{24}\cos\{s(x_2 - x_1)\}\,\mathrm{d}^3r_2\,\mathrm{d}^3r_3\,\mathrm{d}^3r_4. \tag{14}$$

The recurrence equations of the second kind fortunately make it possible to simplify the equations. In fact:

$$\int n_{1234}X_{24}\,\mathrm{d}^3r_4 = kT(\partial n_{123}/\partial x_2) - (X_{21} + X_{23})\,n_{123},$$
$$\int n_{123}X_{13}\,\mathrm{d}^3r_3 = kT(\partial n_{12}/\partial x_1) - X_{12}n_{12},$$
$$\int n_{123}X_{23}\,\mathrm{d}^3r_3 = -kT(\partial n_{12}/\partial x_1) + X_{12}n_{12}.$$

After some transformations we obtain

$$\langle|\mathrm{d}^2B_s/\mathrm{d}t^2|^2\rangle = (s/m)^2\,kT$$
$$\times \left[3s^2kTn - \int(\partial X_{12}/\partial x_1)[1 - \cos\{s(x_2 - x_1)\}]\,n_{12}\,\mathrm{d}^3r_2\right]; \tag{15}$$

and, in the L.W. approximation,

$$\langle|\mathrm{d}^2B_s/\mathrm{d}t^2|^2\rangle = s^4(kT/m^2)$$
$$\times \left[3kTn - \tfrac{1}{2}\int (\partial X_{12}/\partial x_1)(x_2 - x_1)^2\,n_{12}\,\mathrm{d}^3r_2\right]. \tag{16}$$

The analogous quadratic mean relative to transverse momentum waves can be calculated in a similar manner. The result is the

Thermal Waves in Fluids

following, which can be compared with equation (15):

$$\langle |dM_{sy}/dt|^2 \rangle$$
$$= s^2 kT \left[s^2 kTn - \tfrac{1}{2} \int (\partial Y_{12}/\partial y_1) \left[1 - \cos \{ s(x_2 - x_1) \} \right] n_{12} \, d^3 r_2 \right].$$
(17)

8.4. Energy

After momentum it is natural to consider energy. In Section 1.10 we ascribed to each particle a portion of the total energy, using the expression

$$E_J = \tfrac{1}{2} m u_J^2 + \tfrac{1}{2} \sum_K W_{JK}.$$
(1)

It is natural to associate with the set of the E_J an energy wave whose amplitude can be written as

$$E_s = V^{-1/2} \sum_J E_J \exp \left[- i(s \cdot r_J) \right].$$
(2)

Once we have divided the total energy into elements ascribed to each particle, a device which is the better the more uniform the medium is, we should consider only the case of long wavelengths.

The average of E_s is zero. We shall give the calculation of quadratic or mixed averages. The following are the simplest results (L.W.):

$$\langle B_s^* E \rangle = \tfrac{3}{2} kT \left[n + \int (n_{12} - n^2) \, d^3 r_2 \right] + \int n_{12} W_{12} \, d^3 r_2$$
$$+ \tfrac{1}{2} \int (n_{123} - n_{12}n) \, W_{12} \, d^3 r_2 \, d^3 r_3,$$
(3)

$$\langle |E_s|^2 \rangle = \tfrac{1}{4} \left[15(kT)^2 \, n + 9(kT)^2 \int (n_{12} - n^2) \, d^3 r_2 + 12 kT \right.$$
$$\times \int n_{12} W_{12} \, d^3 r_2 + 6 kT \int (n_{123} - n_{12}n) \, W_{12} \, d^3 r_2 \, d^3 r_3$$
$$+ 2 \, n_{12} W_{12}^2 \, d^3 r_2 + 4 \int n_{123} W_{12} W_{13} \, d^3 r_2 \, d^3 r_3$$
$$+ \int (n_{1234} - n_{12}n_{34}) \, W_{12} W_{34} \, d^3 r_2 \, d^3 r_3 \, d^3 r_4 \right],$$
(4)

$$\langle (dB_s^*/dt)(dE_s/dt) \rangle = s^2 (kT/2m)$$
$$\times \left[5 kTn + \int n_{12} W_{12} \, d^3 r_2 - \int n_{12}(x_2 - x_1) X_{12} \, d^3 r_2 \right],$$
(5)

$$\langle M_{sy}(dE_s/dt) \rangle = 0.$$
(6)

The recurrence equations used also in the last section allow us to put (3) in a more condensed form:

$$\langle B_s^* E_s \rangle = \partial \left[\tfrac{3}{2} kTn + \tfrac{1}{2} \int n_{12} W_{12} \, d^3 r_2 \right] / \partial \alpha.$$
(7)

142

8.5. Thermodynamic Interpretation

With the exception of equation (15) of section 8.3, to which we shall return, the results which we obtained for long wavelengths can be expressed with the help of the basic thermodynamic quantities and the variances which relate to them. This gives the following relations:

$$V\langle|B_s|^2\rangle = \langle\Delta N^2\rangle, \tag{1}$$

$$V\langle|dB_s/dt|^2\rangle = kTs^2\langle N\rangle/m, \tag{2}$$

$$V\langle|E_s|^2\rangle = \langle\Delta H^2\rangle, \tag{3}$$

$$V\langle B_s^* E_s\rangle = \langle\Delta N \Delta H\rangle, \tag{4}$$

$$V\langle(dB_s^*/dt)(dE_s/dt)\rangle = kTs^2(U + PV)/m. \tag{5}$$

Formulae (1), (2) and (5) follow immediately. For formulae (3) and (4) one needs an explicit, preliminary calculation of the averages occurring on the right-hand side as functions of the reduced densities and of the intermolecular energy, a calculation which presents no difficulty.

The variances can be eliminated in favour of the derivatives of the basic thermodynamic quantities. We recall the following formulae:

$$\langle\Delta N^2\rangle = \partial\langle N\rangle/\partial\alpha, \tag{6}$$

$$\langle\Delta N \Delta H\rangle = \partial U/\partial\alpha = -\partial\langle N\rangle/\partial\beta, \tag{7}$$

$$\langle\Delta H^2\rangle = -\partial U/\partial\beta. \tag{8}$$

Let us make the significance of these derivatives clear: in the grand canonical distribution the independent variables are α, β and the applied potential. In the case of a uniform medium, which is all that interests us here, the applied potential is replaced simply by the volume. We now propose to substitute for the derivatives (6), (7), (8) more physical coefficients such as the specific heats or the isothermal compressibility coefficient. Let us also recall the following formulae:

$$\ln \Xi = PV/kT, \tag{9}$$

$$\partial \ln \Xi/\partial\alpha = \langle N\rangle, \tag{10}$$

$$\partial \ln \Xi/\partial\beta = -U. \tag{11}$$

An arbitrary infinitesimal transformation can be formulated as follows:

$$dU = \frac{U}{V} dV - \frac{1}{kT^2} (\partial U / \partial \beta)\, dT - (\partial \langle N \rangle / \partial \beta)\, d\alpha, \quad (12)$$

$$d \langle N \rangle = n\, dV - \frac{1}{kT^2} (\partial \langle N \rangle / \partial \beta)\, dT + (\partial \langle N \rangle / \partial \alpha)\, d\alpha, \quad (13)$$

$$dP = [(U + PV)/VT]\, dT + nkT\, d\alpha. \quad (14)$$

Let us first consider an isothermal transformation with constant occupation. We have

$$\partial \langle N \rangle / \partial \alpha = \langle N \rangle\, nkT\chi, \quad (15)$$

χ being the isothermal compressibility coefficient. We have already made use of a result of this kind.

Let us now consider a transformation with constant volume and occupation. We introduce the specific heat at constant volume c by the relation

$$dU = m \langle N \rangle\, c\, dT. \quad (16)$$

We obtain

$$\partial U / \partial \beta + (\partial \langle N \rangle / \partial \beta)^2 / (\partial \langle N \rangle / \partial \alpha) = -m \langle N \rangle\, kT^2 c. \quad (17)$$

This transformation also allows us to make use of the temperature coefficient of the pressure at constant occupation and volume:

$$P_T' = (\partial P / \partial T)_V = (U + PV)/VT + (n/T)\,(\partial \langle N \rangle / \partial \beta)/(\partial \langle N \rangle / \partial \alpha). \quad (18)$$

We therefore have

$$\partial \langle N \rangle / \partial \beta = \langle N \rangle\, kT\chi(TP_T' - P - U/V), \quad (19)$$

$$\partial U / \partial \beta = -\langle N \rangle\, kT[mcT + V\chi(TP_T' - P - U/V)^2]. \quad (20)$$

These relations will be completed by the formula which connects the specific heat at constant pressure C with the specific heat at constant volume:

$$mN(C - c) = V\chi TP_T'^2. \quad (21)$$

Equations (1) to (5) obviously supply us with only incomplete information on the properties of the waves of thermal motion. We should like in particular to have available a "dispersion" law, i.e. a law which will give us the intensity as a function of the frequency. In fact, we have explicit results only for some averages or "moments". We must try to do our best with them.

The simplest idea consists first of all of imagining an approximation where all the waves have the same velocity γ. We could then write

$$B_s = A_s \exp \{i[(s \cdot r) - \omega t]\} \tag{22}$$

with

$$\omega = \pm s\gamma,$$

ω being the angular frequency and A_s being an amplitude independent of time. The alternative of $+$ or $-$ is necessary to allow for the two possible directions of propagation. We assume that the wave of frequency ω is not correlated with the wave of frequency $-\omega$.

This scheme leads us to the following two relations:

$$\langle |B_s|^2 \rangle = \langle |A_s|^2 \rangle \tag{23}$$

and

$$\langle |dB_s/dt|^2 \rangle = \omega^2 \langle |A_s|^2 \rangle. \tag{24}$$

Formulae (1) and (2) supplemented by (6) and (15) then lead to the following result

$$\gamma^2 = (mn\chi)^{-1}. \tag{25}$$

The value of the propagation velocity calculated in this way is the one that the velocity of sound would have if the propagation of sound were an isothermal phenomenon. It is in fact an adiabatic phenomenon. It gives us a shock to find that waves of thermal motion are not propagated, at least to a point, like acoustic waves.

The averages that we have not yet used allow us to overcome this difficulty. But a more complex scheme is necessary. We assume now that for each wavelength there is on the one hand a stationary wave and on the other hand two non-stationary waves with the respective frequencies of ω and $-\omega$ and the same velocity of propagation γ. We shall denote their respective amplitudes by

$$A_0, A_+, A_-,$$

neglecting to specify the suffix s.

Above we were dealing with density waves. We shall formulate the same hypothesis for energy waves which leads us to consider

145

amplitudes $\qquad E_0, E_+$ and E_-.

We shall assume a new approximation by writing a proportionality between the energy amplitudes and the density amplitudes:

$$E_0 = f_0 A_0, \ E_+ = fA_+, \ E_- = fA_-, \tag{26}$$

the coefficients f not being statistical quantities, which means that we are assuming for the calculation of averages relating to the energy, certain relations of which two examples are sufficient:

$$\langle A_0^* E_0 \rangle = f_0 \langle |A_0|^2 \rangle,$$
$$\langle |E_0|^2 \rangle = |f_0|^2 \langle |A_0|^2 \rangle. \tag{27}$$

Lastly we assume the statistical independence of waves of different frequencies

$$\langle A_0^* A_+ \rangle = \langle A_0^* A_- \rangle = \langle A_+^* A_- \rangle = 0 \tag{28}$$

and, naturally, the equality

$$I = \langle |A_+|^2 \rangle = \langle |A_-|^2 \rangle. \tag{29}$$

We shall put

$$I_0 = \langle |A_0|^2 \rangle. \tag{30}$$

Formulae (1) to (5) then lead to the following results:

$$V(I_0 + 2I) = \langle \Delta N^2 \rangle, \tag{31}$$
$$2V\gamma^2 I = kT \langle N \rangle / m, \tag{32}$$
$$V(|f_0|^2 I_0 + 2|f|^2 I) = \langle \Delta H^2 \rangle, \tag{33}$$
$$V(f_0 I_0 + 2fI) = \langle \Delta N \, \Delta H \rangle, \tag{34}$$
$$2V\gamma^2 fI = kT(U + PV)/m. \tag{35}$$

The coefficients f_0 and f are necessarily real. From equations (32) and (35) we can first of all derive the value of the coefficient f:

$$f = (U + PV)/\langle N \rangle. \tag{36}$$

We then have

$$f_0 = \frac{\langle \Delta H^2 \rangle - f \langle \Delta N \, \Delta H \rangle}{\langle \Delta N \, \Delta H \rangle - f \langle \Delta N^2 \rangle}, \tag{37}$$

$$VI_0 = \frac{(\langle \Delta N \, \Delta H \rangle - f \langle \Delta N^2 \rangle)^2}{\langle \Delta H^2 \rangle + f^2 \langle \Delta N^2 \rangle - 2f \langle \Delta N \, \Delta H \rangle},$$

$$2VI = \frac{\langle \Delta N^2 \rangle \langle \Delta H^2 \rangle - \langle \Delta N \, \Delta E \rangle^2}{\langle \Delta H^2 \rangle + f^2 \langle \Delta N^2 \rangle - 2f \langle \Delta N \, \Delta H \rangle},$$

$$\gamma^2 = nkT/2mI.$$

146

The complicated quantities which still occur can be expressed simply in terms of the specific heats and of the isothermal compressibility coefficient. We obtain

$$2I = \frac{c}{C} n^2 kT\chi, \tag{38}$$

$$I_0 = \frac{C - c}{C} n^2 kT\chi, \tag{39}$$

$$\gamma^2 = \frac{C}{c} (mn\chi)^{-1}. \tag{40}$$

We finally have

$$f_0 = \frac{U + PV}{\langle N \rangle} - \frac{C}{C - c} \frac{TP'_T}{n}. \tag{41}$$

Our theoretical scheme has supplied us with a number of unknowns exactly adjusted to the number of equations available. The solution has presented no difficulty. The calculation gave the respective intensities of the non-stationary waves [equation (38)], of the stationary waves [equation (39)], and the velocity of propagation of the non-stationary waves [equation (40)]. This velocity is the same as the velocity of sound which results from the elementary theory.

This set of results has been known since the work of Landau and Placzek (see, e.g., Gross, 1945). They were obtained by means of macroscopic considerations. Microscopic theory shows which approximations are necessary. The reader will note that it has been necessary to back up the data relating to the pure density waves by data on the energy waves. Or, again, that kinetics by itself was not sufficient and that it had to be supplemented by hydrodynamic and thermodynamic concepts.

A result of this theory is that monochromatic light which travels through a liquid gives rise to a complex scattered light. The Rayleigh ray has the structure of a triplet: the central ray has exactly the same wavelength as the incident ray. It is due to scattering by stationary waves. The other rays are caused by acoustic waves. The frequency shift is due to the same principle as the Doppler and Fizeau effect. The two displaced rays have the same intensity and the ratio of the intensity of the whole of the doublet to the total scattered intensity is equal to c/C. The above statements are obviously not an attempt to go more deeply into the phenomena from the point of view of optics.

Experimental verification has long remained difficult, since monatomic liquids, which are the only convenient ones, scatter light too little. Transparent single crystals, for which the theory is very similar in many points, are most suitable for observation.†
The invention of the laser is in the process of bringing the whole subject up again.

It is not a question that can be answered today. Moreover, the next section will bring up the limits of the theory of thermal waves which has just been given.

8.6. Higher-order Moments

The above analysis has discreetly forgotten about several averages of which, however, we gave a first estimate in section 3.8. Let us in particular take a further look at the following result (L.W.):

$$\langle |d^2 B_s / dt^2|^2 \rangle = s^4 (kT/m^2)$$
$$\times \left[3nkT - \tfrac{1}{2} \int n_{12}(\partial X_{12}/\partial x_1)(x_2 - x_1)^2 \, d^3 r_2 \right]. \tag{1}$$

The above average is not reducible to thermodynamic quantities which can be exactly determined experimentally. It is possible nevertheless to give an estimate for it.

The problem is to transform the integral term. Let us first of all note the following equality:

$$-\frac{1}{2} \int n_{12}(\partial X_{12}/\partial x_1)(x_2 - x_1)^2 \, d^3 r_2$$
$$= \frac{1}{15} \int n_{12}r(d W_{12}/dr) \, d^3 r_2 + \frac{1}{10} \int n_{12}r^2(d^2 W_{12}/dr^2) \, d^3 r_2. \tag{2}$$

The first term on the right-hand side is connected with the pressure which obtains in the fluid in equilibrium:

$$\int n_{12}r(d W_{12}/dr) \, d^3 r_2 = 6(nkT - P). \tag{3}$$

The second term can only be estimated, at least if we assume as an acceptable approximation a force law of the following form:

$$W_{12} = \varepsilon[(r/r_0)^{-\mu} - (r/r_0)^{-\nu}], \tag{4}$$

† See Rousset (1947). The scattering of neutrons, even of those which are the slowest that we can observe, does not correspond to the range of long wavelengths which is all we have gone into here.

because then the term in question can be expressed as a function of the pressure and of the internal energy. The force law (4) is a generalization of the Lennard-Jones law. Simple calculations lead to the following result:

$$\int n_{12} r^2 (d^2 W_{12}/dr^2) \, d^3 r_2 = 6(\mu + \nu + 1) (P - nkT)$$
$$- \mu\nu(2U/V - 3nkT). \tag{5}$$

If we admit this calculation the relation (1) becomes

$$\langle |d^2 B_s/dt^2|^2 \rangle = \frac{1}{10} s^4 \frac{kT}{m^2}$$
$$\times [-2\mu\nu U/V + (28 + 3\mu\nu - 6\mu - 6\nu) \, nkT + (2 + 6\mu + 6\nu) \, P]. \tag{6}$$

For a liquid observed under atmospheric pressure and far from its critical point the principal term is the one that corresponds to the internal energy. Let us adopt Lennard-Jones's values:

$$\mu = 12, \qquad \nu = 6.$$

The following formula which contains L, the heat of vaporization per unit of mass at constant pressure in energy units, gives a sufficiently approximate version of equation (6) for discussion:

$$\langle |d^2 B_s/dt^2|^2 \rangle \sim 14 s^4 \, nkTL/m. \tag{7}$$

On the other hand, the scheme suggested in the preceding section leads to the following relation:

$$\langle |d^2 B_s/dt^2|^2 \rangle = 2\omega^4 I. \tag{8}$$

Published data show that the estimate (7) leads to a result which is two or three times too great. The scheme is too simple.

Leaving this point aside—it is the subject of new research—we now propose to show that the average which we have just studied is in a certain way connected with pressure fluctuations. In this book the desire to present pressure as a quantity that is defined from point to point, so as to satisfy the needs of hydrodynamics, has led us to neglect the concept of the virial. Without going into useless details let us consider the following physical quantity

$$\mathscr{P}_{xy} = \left[\sum_J p_{Jx} p_{Jy}/m - \tfrac{1}{2} \sum_J \sum_K (y_K - y_J) X_{JK} \right]/V, \tag{9}$$

which is nothing other than the time derivative of the quantity

$$\sum_J y_J p_{Jx}/V.$$

This is a symmetrical tensor whose average—in a uniform medium—is the same as the pressure tensor. Calculations of the type appearing in this chapter, details of which we shall spare the reader, allow us to show that the average relating to the density waves that we have been studying in the present section is connected with the variance of $\mathscr{P}$:

$$m^2 \langle |d^2 B_s/dt^2|^2 \rangle = V s^4 \langle |\mathscr{P}_{xx} - P|^2 \rangle. \tag{10}$$

This result is not valid—in this form at least—except for long wavelengths.

Equation (10), which is interesting in itself, becomes more important when we are dealing, as we are going to do now, with a new average which was missing in the picture of our averages relating to the energy and which, however, is the natural complement of it. It is a question of the following average

$$\langle |dE_s/dt|^2 \rangle \tag{11}$$

of which a direct calculation gives the following expression (L.W.):

$$\langle |dE_s/dt|^2 \rangle = \tfrac{1}{4} s^2 (kT/m) \Big[35n(kT)^2$$
$$+ 10kT \int [W_{12} - X_{12}(x_2 - x_1)]\, n_{12}\, d^3 r_2$$
$$+ \int [W_{12}^2 + \{X_{12}^2\}(x_2 - x_1)^2 - 2W_{12}X_{12}(x_2 - x_1)]\, n_{12}\, d^3 r_2$$
$$+ \int [W_{12}W_{13} + \{X_{12}X_{13}\}(x_2 - x_1)(x_3 - x_1)$$
$$- 2W_{12}X_{13}(x_3 - x_1)]\, n_{123}\, d^3 r_2\, d^3 r_3 \Big]. \tag{12}$$

This average has a point in common with the average (10): it cannot be reduced to simple thermodynamic quantities. The least that can be done is to give a physical interpretation of it similar to the one provided by formula (10). For this purpose we shall introduce a quantity which will here play the part that the tensor $\mathscr{P}_{xy}$ played in connexion with the pressure.

The parallelism which exists between momentum and energy leads us to consider here the time derivative of the quantity

$$\sum_J x_J E_J/V,$$

or

$$\mathscr{C}_x = \frac{\sum_J p_J\{p_J^2\}/m + \sum_J \sum_K p_J W_{JK} - \frac{1}{2}(x_K - x_J)\{(p_s + p_K) X_{JK}\}}{2mV}.$$

(13)

This vector, of which we have already met a local average in the study of energy diffusion—formula (36) of section 1.8—plays the same part in connexion with the heat flux as the tensor $\mathscr{P}_{xy}$ plays with respect to the pressure, except that its average is zero in thermal equilibrium.

Once we know this, we see that the average (11) is closely linked with the variance of the vector $\mathscr{C}_x$:

$$\langle |dE_s/dt|^2 \rangle = Vs^2 \langle \mathscr{C}_x^2 \rangle. \tag{14}$$

Equations (10) and (14) are physically interesting but their practical consequences are not obvious.

In thermal waves we have been tempted to see not genuinely periodic waves but waves damped by the effect of viscosity and thermal conductivity. This picture would be satisfactory if we could produce, as regards the sources of the damping, sources which would be responsible for constantly restoring them since thermal motion is a permanent phenomenon. This problem has not been solved. The present section will at least have achieved its aim if it puts us on our guard against incomplete interpretations.

CHAPTER 9

Entropy and Heat

9.1. Introduction

The purpose of this chapter is to study the relations which exist between entropy and heat.

The transfer of energy from one part of a system to another part of the same system through heat conduction belongs to the study of non-equilibrium conditions. It should be stated, so as to put ourselves in a clearly defined situation, that the system is isolated. In this way certain phenomena which have the appearance of permanent conditions are included in the category of non-equilibrium phenomena: this appearance is due to the fact that attention has been paid only to an unisolated fragment of a larger system.

These phenomena involving heat are sometimes reversible, but more generally they are irreversible. The theory of irreversible phenomena, in which reversible phenomena have been included as the limiting case, is not in the programme of the present book. We shall only allude to it here. Taken in a completely statistical framework this theory presents great difficulties. We shall stick here to the case of moderately dense gases and to the case of short-range forces: we have available there the body of doctrine which we owe to Boltzmann, and particularly his famous H-theorem. We shall try and place this in the framework of our study of statistical entropy.†

9.2. Evolution of Statistical Entropy of an Isolated System

We have already mentioned the fact that the statistical entropy of an isolated system is independent of the time. The reason for

† The problems that occur have already been dealt with by Grad (1961).

152

this is that the density in phase satisfies the Liouville equation. This argument, which is valid for a system whose occupation is fixed, becomes generalized when it is not.

It is within the framework of a situation of this kind that we shall be continuing our investigations. The advantage of this is that we can sometimes invoke the short range of the correlations. We are considering a system which contains particles of only one kind. The calculations in the entropy chapter now allow us to put this in the following form:

$$k^{-1}S = \int \mu_1(1 - \ln \Gamma\mu_1) \, \mathrm{d}\Omega_1 + \tfrac{1}{2}\int (- \mu_{12} \ln a_{12} + \mu_1\mu_2\varepsilon_{12})$$
$$\times \, \mathrm{d}\Omega_{12} + \cdots. \tag{1}$$

Here we have a kind of expansion in rising powers of the density, but whose convergence risks being uncertain if the range of the correlations is long. Only the first two terms have been written down explicitly because we intend to treat gases only.

Let us return to the calculation of the evolution of the entropy. We use the expansion (1) and the chain of recurrence equations of motion. Let us rewrite the first recurrence equation:

$$\left(\frac{\partial}{\partial t} + \frac{p_1}{m}\frac{\partial}{\partial x_1} + X_1 \frac{\partial}{\partial p_1}\right)\mu_1 + \int X_{12}\frac{\partial}{\partial p_1}\mu_1\mu_2 a_{12} \, \mathrm{d}\Omega_2 = 0. \tag{2}$$

Bearing in mind the invariance of the occupation expressed by the relation

$$(\mathrm{d}/\mathrm{d}t)\int \mu_1 \, \mathrm{d}\Omega_1 = 0,$$

the time derivative of the first-order term in the entropy is:

$$\frac{\mathrm{d}}{\mathrm{d}t}\int \mu_1(1 - \ln \Gamma\mu_1) \, \mathrm{d}\Omega_1 = - \int (\partial\mu_1/\partial t) \ln \mu_1 \, \mathrm{d}\Omega_1. \tag{3}$$

We take $\partial\mu_1/\partial t$ from formula (2). The conditions at the limits make all the first-order terms disappear. After some transformations we have

$$\frac{\mathrm{d}}{\mathrm{d}t}\int \mu_1(1 - \ln \Gamma\mu_1) \, \mathrm{d}\Omega_1 = - \int (\partial\mu_1/\partial p_1) X_{12}\mu_2 a_{12} \, \mathrm{d}\Omega_{12}. \tag{4}$$

In general this quantity is not zero. We must expect each of the terms of the expansion (1) to evolve. Only their sum can be invariant.

Using the second, and also the first, recurrence equation, we can similarly calculate the derivative of the second-order term. The results include a second-order quantity and a third-order quantity. The second-order quantity is:

$$\int (\partial\mu_1/\partial p_1) \, X_{12}\mu_2 a_{12} \, d\Omega_{12};$$

it compensates exactly for the second-order quantity which we found for the derivative of the first-order term.

The sum of the first two terms of the entropy therefore has a third-order derivative.

This result must be considered to be a general one. The derivative of the sum of the first n terms of the entropy, a sum which expresses the entropy up to the order of n inclusive, comprises only a quantity of the order of $n + 1$. The result is, provided that the expansion is convergent, that this expansion has a sum that is independent of time. This result merely confirms the one that was obtained directly by proceeding from the Liouville equation. This property is entirely foreign to the entropy of thermodynamics.

We have assumed the system to be isolated. This point implies that the applied field is independent of time. In fact, the above calculations which are based either on the Liouville equation or on the recurrence equations of motion are not modified when the applied field is time-dependent. In the following sections, however, we must keep to the case of a rigorously isolated system.

9.3. The *H*-Theorem

The *H*-theorem relates to gases and short-range forces. The density is such that two molecules are quite likely to be in contact with each other, but it is such that the meeting of three molecules or more is highly improbable. This possibility will therefore be neglected in the calculations.

The collision of two molecules is generally a fairly brief phenomenon which can be described by using the mechanics of systems of two point particles. We take no notice of the appearance that other molecules may make and for the duration of this phenomenon we neglect the action which the applied field could have: this last approximation is justified if the applied field is moderate, which is generally the case a long way from the walls, where intermolecular interactions definitely predominate. At the wall the applied field cannot be treated so lightly: we shall content ourselves with assuming that the formalism which we are going to develop for the regions where the field is weak could be generalized, with its symmetry characteristics, when it is strong.

The collision comprises three stages:

(1) the two molecules approach each other;
(2) the effect of the forces begins to make itself felt and modifies the trajectories little by little. This modification may be considerable if the molecules have a hard core;
(3) the two molecules separate.

Each collision is characterized by the relative velocity of the two participants and by the impact parameter, which is the minimum distance to within which the two molecular centres would come if there were no force exerted between the molecules in question.

The hypothesis of molecular chaos allows the effect of the collisions to be calculated: it consists of assuming that there is no correlation between two molecules that are going to collide in the time that precedes the collision.

The result is that there are correlations during the collision and after the collision.

The molecular chaos hypothesis has a rather unsatisfactory aspect in that this hypothesis is called on at each point in time. There is no question in Boltzmann's argument of stating it for a certain point in time, that could be called the initial point in time, and of showing that from then on the property is preserved indefinitely. It should be noted in passing that, to be consistent and because we cannot see why the initial point in time should be privileged, we should be able to extend the validity of the hypothesis back to points in time earlier than the initial point in time.†

It is quite possible that the difficulties implied by the molecular chaos hypothesis are due to the fact that underlying its statement are two propositions, one of which would be a true hypothesis and the other of which would only be an approximation justified in the case of dilute media. The existence of correlations during collision

† The idea of pushing the initial point in time back indefinitely brings up questions which are beyond the present possibilities of observations. See a remark by ter Haar (1954, p. 335).

is the more understandable since they appear in thermal equilibrium.

The correlations appear after the collision: it is probably reasonable to assume that they are "diluted" by successive collisions, which is not exactly assuming that there are no correlations at all before a collision, but that the correlation level is very small when compared with that engendered by the collision itself.

I do not intend to go any deeper into this delicate problem here. But, bearing in mind the hypothesis of molecular chaos, what I propose to show is that the hypothesis in question does not contradict as seriously as might be feared the theorem of the invariance of the statistical entropy of an isolated system.

Boltzmann derived from the hypothesis of molecular chaos the equation which bears his name and which can be written as follows:

$$\left(\frac{\partial}{\partial t} + \frac{p_1}{m}\frac{\partial}{\partial x_1} + X_1\frac{\partial}{\partial p_1}\right)\mu_1$$

$$= \frac{1}{m}\int (\mu_1'\mu_2' - \mu_1\mu_2)|p_2 - p_1|\, b\, \mathrm{d}b\, \mathrm{d}\varphi\, \mathrm{d}^3r_2. \tag{1}$$

The second term is what is called the collision term or, when the stress is on the hard core, the impact term. With a few minor approximations it is nothing other than the interaction term of the first recurrence equation, with opposite sign, as it becomes when into the general formalism we introduce the hypothesis of molecular chaos; b is the impact parameter and φ is a polar angle which places the virtual point of minimum approach in a plane normal to the relative velocity. All the densities μ_1, μ_2, μ_1', μ_2' are taken at the same point r_1, at the same time t; they differ by the momentum they refer to:

p_1 for the molecule hit, i.e. the only one that occurs on the left-hand side,

p_2 for the colliding molecule,

p_1' and p_2', respectively, for the same molecules after the collision.

Mathematical use of the Boltzmann equation is made inconvenient by the complicated dependence of p_1' and of p_2' on p_1, p_2 and b which appear in the collision term as independent variables and by the fact that it is not linear.

From the Boltzmann equation we can deduce that the entropy

$$S_1 = k \int \mu_1(1 - \ln \Gamma\mu_1)\, \mathrm{d}\Omega_1 \tag{2}$$

increases in the course of time. As the quantity

$$\int \mu_1 \, d\Omega_1$$

is invariant, a general property which the Boltzmann equation conserves, it is sufficient to show that the quantity

$$H = \int \mu_1 \ln \mu_1 \, d\Omega_1 \tag{3}$$

decreases. Let us find the derivative of this quantity from the Boltzmann equation. We obtain

$$\frac{d}{dt} H = \frac{1}{m} \int \ln \mu_1 (\mu_1' \mu_2' - \mu_1 \mu_2) |p_2 - p_1|$$
$$\times b \, db \, d\varphi \, d^3r_1 \, d^3p_1 \, d^3p_2. \tag{4}$$

Nothing is changed in the result if we permute the momenta p_1 and p_2. Equation (4) can therefore equally well be written as

$$\frac{dH}{dt} = \frac{1}{2m} \int \ln (\mu_1 \mu_2) (\mu_1' \mu_2' - \mu_1 \mu_2) |p_2 - p_1|$$
$$\times b \, db \, d\varphi \, d^3r_1 \, d^3p_1 \, d^3p_2. \tag{5}$$

Let us now associate with the collision characterized by the initial conditions p_1, p_2, b and finishing with p_1', p_2' with an "exit parameter" b' the collision that starts with p_1', p_2', b' and which finishes with p_1, p_2, b. By virtue of the mechanics of the relative motion we have the following relations available to us:

$$b = b', \quad d\varphi = d\varphi', \quad |p_2' - p_1'| = |p_2 - p_1|, \tag{6}$$

to which the Liouville theorem adds

$$d^3p_1 \, d^3p_2 = d^3p'_1 \, d^3p'_2.$$

We now obtain

$$\frac{dH}{dt} = \frac{1}{4m} \int \ln \frac{\mu_1 \mu_2}{\mu_1' \mu_2'} (\mu_1' \mu_2' - \mu_1 \mu_2) |p_2 - p_1|$$
$$\times b \, db \, d\varphi \, d^3r_1 \, d^3p_1 \, d^3p_2. \tag{7}$$

The quantity

$$(\mu_1' \mu_2' - \mu_1 \mu_2) \ln (\mu_1 \mu_2 / \mu_1' \mu_2') \tag{8}$$

can only be negative or zero. From this results the H-theorem which can be stated as follows: the entropy term S_I can only increase or remain stationary.

Stationariness obtains when the following equality is satisfied:

$$\mu_1' \mu_2' = \mu_1 \mu_2. \tag{9}$$

The quantity $\mu_1 \mu_2$ must be invariant during the collision. It is therefore a function of the latter's invariants, at least of those which depend only on the momentum, namely

$$p_1 + p_2 \quad \text{and} \quad (p_1^2 + p_2^2)/2m. \tag{10}$$

The logarithm of the simple density in phase is therefore necessarily a linear function of the momentum and of the kinetic energy, which we can express again by the following formula:

$$\mu_1 = A_1 \exp\left[-\{(a_1 \cdot p_1) + \beta_1 p_1^2/2m\}\right], \tag{11}$$

where the factors A_1, β_1 and the vector a_1 can depend on spatial coordinates. At the same time the collision term of the Boltzmann equation is zero. Let us then substitute the expression (11) in what remains of equation (1). We obtain

$$\left(\frac{\partial}{\partial t} + \frac{p_1}{m}\frac{\partial}{\partial x_1} + X_1\frac{\partial}{\partial p_1}\right) \times (\ln A_1 - (a_1 \cdot p_1) - \beta_1 p_1^2/2m) = 0. \tag{12}$$

This equation must be discussed allowing for our usual conditions at the limits: the wall is essentially a repulsive potential and the molecules can never reach its "hard" limit.

Two cases occur according to whether the applied field, including the one that represents the wall, has spherical symmetry or not.

In the first case β is necessarily uniform and independent of time, the vector a is zero and the density satisfies the relation:

$$kT(\partial n_1/\partial x_1) = X_1 n_1. \tag{13}$$

This is the equation for the distribution of a perfect gas in thermal equilibrium in an applied field.

In the second case the equilibrium does not have to obtain because we can substitute an overall motion which is a rotation similar to the rotation of a solid round an axis passing through the system's centre of symmetry.

To describe the usual physical situation in the best possible way it must be assumed that the wall has a microscopic grain: the wall repels the molecules which collide with it in a random manner.†
We have already mentioned the usefulness of a structure of this

† This type of wall has been studied by Rocard (1932).

kind in connexion with the theory of surface tension. Under these conditions the case of perfect spherical symmetry is of no physical interest.

Let us note in any case that perfect spherical symmetry of the field could be achieved in a system of free particles which exert gravitational forces onto one another. But it would then be a question of a self-consistent field and not of an applied field: the Boltzmann equation must be manipulated for the treatment of this kind of problem. We are not stressing this here.

The Boltzmann equation is nothing other than the first recurrence equation where the two by two correlations have been given a particular expression. We should expect to obtain as a result a more precise equation than (13) for formulating the density distribution in equilibrium. Equation (13) is equivalent to the historic Laplace equation. A more complete equation would be expected here. This equation would be equivalent to the generalized Laplace equation, which would have had all terms removed of order higher than two in the density. The reason for this shortcoming is that the medium was assumed to be uniform on the molecular scale in order to obtain the Boltzmann equation.

We shall assume that a more rigorous way of writing the consequences of the molecular chaos hypothesis would alter nothing in the *H*-theorem at the same time as providing for the description of the equilibrium a completely correct relation and not equation (13).

Having accepted this point we can put forward the following interpretation of the *H*-theorem: it is not the traditional interpretation. In our approximation of the moderately dense gas the statistical entropy is composed of two terms:

$$S = S_I + S_{II}.$$

The *H*-theorem shows us that S_I can grow only in the course of time. Our general arguments teach us that the total entropy is a constant. There is no contradiction here: S_{II} decreases as S_I increases.

This interpretation alters nothing in the other consequences of the *H*-theorem which can be summed up in one sentence: the system becomes thermalized a little at a time.

It is better to use the concept of thermalization here than that of thermodynamic equilibrium because the probabilities of occupation (the parameters π_N) do not evolve: they cannot therefore adjust themselves exactly to the requirements of the grand canonical thermodynamic equilibrium. The same is also true of the energy distribution law which does not evolve either.

These difficulties are not enough to make us refrain from using the idea of a poorly determined occupation, or the idea of a certain arbitrariness in the energy distribution: a strict occupation and a well-defined energy—i.e. the microcanonical distribution—are physical illusions and mathematically conceal traps which it is preferable to avoid.

The grand-canonical thermodynamic equilibrium with all its attributes can occur only if we excise a much smaller partial system from the isolated system: the latter will contain on the average only a small part of the energy and of the available particles.

9.4. Statistical Entropy Density

This concept is convenient for studying the changes in entropy that we shall touch on at the end of this section.

We are still considering a system of identical particles. It is and remains of moderate density during the phenomena which we describe so that the entropy can be limited to its first two terms. The entropy of a partial volume B is by definition given by the following expression:

$$k^{-1}S_B = \int_B \mu_1(1 - \ln \Gamma\mu_1) \, d\Omega_1$$
$$+ \tfrac{1}{2} \int_B \int_B (- \mu_{12} \ln a_{12} + \mu_1\mu_2\varepsilon_{12}) \, d\Omega_1 \, d\Omega_2, \tag{1}$$

the integrations in space being limited to the volume B. A similar expression is valid for the remaining volume C.

The entropy of the total system A is slightly less than the sum of the entropies S_B and S_C: nevertheless the correlation entropy between the volumes B and C is a generally negligible surface effect. For the sake of convenience we shall ascribe to our two partial volumes respectively a part of the correlation entropy. The entropies defined by equation (1) will be replaced by quantities that are only very slightly different but are such that we have exactly

$$S_A = S_B + S_C. \tag{2}$$

We shall therefore have to substitute the following formula for formula (1):

$$k^{-1}S_B = \int_B \mu_1(1 - \ln \Gamma\mu_1)\, d\Omega_1$$

$$+ \tfrac{1}{2}\int_B \int_A (-\mu_{12} \ln a_{12} + \mu_1\mu_2\varepsilon_{12})\, d\Omega_1\, d\Omega_2. \tag{3}$$

This convention resembles that which we adopted for the energy (in Chapter 1) when we needed to share the intermolecular energy between two adjacent systems. Formula (3) now allows us to define a statistical entropy density s_1 by the following relation:

$$k^{-1}s_1 = \int \mu_1(1 - \ln \Gamma\mu_1)\, d^3p_1$$

$$+ \tfrac{1}{2}\int \int_A (\mu_{12} \ln a_{12} + \mu_1\mu_2\varepsilon_{12})\, d^3p_1\, d\Omega_2, \tag{4}$$

this quantity being a function of a set of space and time coordinates. The entropy of the volume B (new kind) can be derived from the entropy density by a simple integration

$$S_B = \int_B s_1\, d^3r_1. \tag{5}$$

Let us study the evolution of s_1. The study of the first term is sufficiently characteristic for it to be unnecessary to begin with to encumber ourselves with the second, unlike what happens for the H-theorem.

Accordingly it will be sufficient for us to write the first recurrence equation to first order only. We have

$$\frac{\partial\mu_1}{\partial t} + \left(\frac{p_1}{m}\frac{\partial}{\partial x_1} + X_1\frac{\partial}{\partial p_1}\right)\mu_1 = 0, \tag{6}$$

or,

$$\frac{\partial}{\partial t}\mu_1(1 - \ln \Gamma\mu_1) = -\left(\frac{p_1}{m}\frac{\partial}{\partial x_1} + X_1\frac{\partial}{\partial p_1}\right)\mu_1(1 - \ln \Gamma\mu_1). \tag{7}$$

(We are not considering the case of an ionized medium which would, as in the case of the energy, introduce some complications here.)

Consequently

$$k^{-1}\frac{\partial s_1}{\partial t} = -\int\left(\frac{p_1}{m}\frac{\partial}{\partial x_1} + X_1\frac{\partial}{\partial p_1}\right)\mu_1(1 - \ln \Gamma\mu_1)\, d^3p_1. \tag{8}$$

Entropy and Heat

This expression can immediately be simplified and reduced to the following:

$$k^{-1}(\partial s_1/\partial t) = -(\partial/\partial x_1) \int (p_1/m)\,\mu_1(1 - \ln \Gamma\mu_1)\,d^3 p_1. \tag{9}$$

Let us single out in the second term the average momentum $\langle p_1 \rangle$ or the average velocity

$$\langle c_1 \rangle = \langle p_1 \rangle/m.$$

We obtain

$$\frac{\partial s_1}{\partial t} = -\frac{\partial}{\partial x_1} \langle u_1 \rangle\, s_1 - \frac{\partial}{\partial x_1} \int \frac{P_1}{m} \mu_1(1 - \ln \Gamma\mu_1)\,d^3 p_1. \tag{10}$$

On the right-hand side a first term appears which expresses the dragging along of the entropy by the average motion and a second term which brings in the fluctuation of the momentum.

Let us recall the definition of $\langle p_1 \rangle$. We put first of all

$$\mu_1 = n_1 f_1.$$

f, normalized to unity, is the momentum distribution function. We then have:

$$\langle p_1 \rangle = \int p_1 f_1\,d^3 p_1.$$

In the same way an average kinetic energy is defined at each point †

$$\frac{1}{2m} \langle P_1^2 + Q_1^2 + R_1^2 \rangle = \frac{1}{2m} \int (P_1^2 + Q_1^2 + R_1^2) f_1\,d^3 p_1. \tag{11}$$

Proceeding from this quantity we shall define at each point a temperature T_1:

$$\frac{3}{2} kT_1 = \frac{1}{2m} \langle P_1^2 + Q_1^2 + R_1^2 \rangle. \tag{12}$$

This temperature will allow us to put the momentum distribution function in the following form:

$$f_1 = \frac{1}{(2\pi mkT_1)^{3/2}} \exp\left[-\frac{P_1^2 + Q_1^2 + R_1^2}{2mkT_1}\right](1 + \varphi_1), \tag{13}$$

the quantity φ_1 being there to express the fact that in general the momenta, or the velocities, are not thermalized. We shall denote

---

† In what follows we are not using summation over repeated indices (see Note on Notation on p. xix).

162

the Maxwell factor by F_1. In our notation it is clear that we have

$$\int F_1 \varphi_1 \, d^3p_1 = 0, \tag{14}$$

$$\int P_1 F_1 \varphi_1 \, d^3p_1 = 0, \tag{15}$$

$$\int (P_1^2 + Q_1^2 + R_1^2) F_1 \varphi_1 \, d^3p_1 = 0, \tag{16}$$

$$\subset P_1^2 \supset - mkT_1 = \int P_1^2 F_1 \varphi_1 \, d^3p_1, \tag{17}$$

$$\subset P_1 Q_1 \supset = \int P_1 Q_1 F_1 \varphi_1 \, d^3p_1, \tag{18}$$

$$C_{1x} = \frac{n}{2m^2} \subset P_1(P_1^2 + Q_1^2 + R_1^2) \supset$$

$$= \frac{n}{2m^2} \int P_1(P_1^2 + Q_1^2 + R_1^2) F_1 \varphi_1 \, d^3p_1. \tag{19}$$

The vector C_1 is the heat flux vector in the first-order approximation. The fact that the kinetic pressure tensor is not that of a perfect gas in equilibrium and the fact that the heat flux vector is not zero are due to the deviations from thermalization.

Let us introduce the expression (13) into the logarithm which appears in (10). We obtain, omitting the suffices,

$$\frac{\partial s}{\partial t} + \frac{\partial}{\partial x} \subset u \supset s + \frac{\partial}{\partial y} \subset v \supset s + \frac{\partial}{\partial z} \subset w \supset s$$

$$= - k \int \frac{1}{m} \left(\frac{\partial}{\partial x} P + \frac{\partial}{\partial y} Q + \frac{\partial}{\partial z} R \right) \mu$$

$$\times \left[1 + \frac{P^2 + Q^2 + R^2}{2mkT} - \ln \frac{\Gamma}{(2\pi mkT)^{3/2}} - \ln(1 + \varphi) \right] d^3p,$$
$$\tag{20}$$

or

$$\frac{\partial s}{\partial t} + \frac{\partial}{\partial x} \subset u \supset s + \frac{\partial}{\partial y} \subset v \supset s + \frac{\partial}{\partial z} \subset w \supset s$$

$$= - \operatorname{div} \frac{C}{T} + \frac{1}{m} k \operatorname{div} \int PnF(1 + \varphi) \ln(1 + \varphi) \, d^3p. \tag{21}$$

The statistical entropy contained in a volume whose boundary follows the hydrodynamic motion is subject to variations which are equal to the flux of the following vector:

$$- \frac{C}{T} + \frac{k}{m} \int PnF(1 + \varphi) \ln(1 + \varphi) \, d^3p. \tag{22}$$

If at a certain point thermalization is nearly realized we can consider that φ is an infinitesimal. Bearing in mind relation (15), we shall therefore be justified in replacing in the last integral

$$(1 + \varphi) \ln (1 + \varphi) \quad \text{by} \quad \tfrac{1}{2}\varphi^2. \tag{23}$$

The heat flux itself is linear in φ. There will therefore be situations when the heat flux alone will be responsible for the variations in the statistical entropy. These variations will be in accordance with the overall formula

$$\Delta S = \Delta\sum(Q/T), \tag{24}$$

Q being the incoming heat. For this result to be applied it is necessary that the thermalization should obtain in practice everywhere on the boundary of the volume in question. On the other hand, there is no condition on what happens in the interior, where the density may have any value.

It should not be forgotten that our arguments relate only to low densities at the boundary. They are nevertheless an attempt at a first approach to a general result.

9.5. Thermal Entropy

The principal result of the last section looks formally like the Carnot principle, but different statements are in fact involved. The equivalence between formula (24) of section 9.4 and the Carnot principle can be considered only when the transformation, occurring in the system studied, is reversible not only at the system's boundary but also at each point in the volume it occupies.

The statistical entropy is not the best quantity to consider if we wish to obtain an interpretation of the Carnot principle in molecular theory. We shall achieve our object by defining a new entropy which we shall call "thermal entropy".

In thermodynamic equilibrium the situation is simple: statistical entropy and thermal entropy are merely the same quantity, i.e. thermodynamic entropy.

We shall borrow the definition we need from the phenomenological theory of irreversible phenomena (see de Groot and Mazur, 1962). We shall not make many claims and, as in the last section, we shall consider only the case of low densities. We give only an outline here which will not even touch on the problems posed by the correlations.

In expanding the statistical entropy or more precisely in expanding its density the general idea is to substitute for the real densities in phase the densities in phase which would obtain around the point in question if the medium were in thermal equilibrium, this thermal equilibrium being adjusted so as to respect the true molecular density and the true energy density. A complete programme results from this idea which phenomenological theory solves without difficulty when the whole system is almost in thermodynamic equilibrium from point to point, but which should be explored in all its generality. The modest limits which we have set ourselves will free us from as difficult an investigation as this.

The case of the perfect gas is fairly clear-cut. At each point there is a molecular density and a temperature (the latter has been defined in the last section). If the medium were Maxwellian the statistical entropy density would be given by the following expression:

$$k^{-1}s' = \int nF(1 - \ln \Gamma nF)\, d^3\boldsymbol{p}. \tag{1}$$

Here again the contribution of the correlations is neglected. It is the quantity s' which will, in the approximation being used, be by definition the thermal entropy density. It clearly depends only on the local molecular density and temperature. It will be noted that here the given temperature and the given energy density are immediately equivalent because in our approximation the energy is reduced to the kinetic energy. The expression (1) can be evaluated to give

$$s' = kn\{\tfrac{5}{2} + \ln [(2\pi mkT)^{3/2}/\Gamma n]\}. \tag{2}$$

The difference between the thermal entropy and the statistical entropy is connected with the existence of the quantity φ:

$$k^{-1}(s' - s) = n \int F(1 + \varphi) \ln (1 + \varphi)\, d^3\boldsymbol{p}. \tag{3}$$

When φ is small this difference is of the second order in this quantity and positive. This last result is, moreover, general:

$$s' \geqq s. \tag{4}$$

The equality holds only when φ is zero.

Let us study the time variation of the thermal entropy. First of all, in accordance with formula (2) we have

$$ds' = (s' - kn)\, dn/n + \tfrac{3}{2} kn\, dT/T. \tag{5}$$

Entropy and Heat

The time derivatives of the density and the temperature can be derived from the general diffusion equations. We use the derivatives along the hydrodynamic trajectory and return to tensor notations, i.e. summation over repeated indices (see Note on Notation, p. xix):

$$dn/dt = -n\partial\langle u\rangle/\partial x, \tag{6}$$

$$mn\,d\langle u\rangle/dt = nX - \partial P_{xy}/\partial y, \tag{7}$$

$$n\,d\langle H\rangle/dt = -\partial(\langle u\rangle P_{xy})/\partial y - \partial C_x/\partial x. \tag{8}$$

Here the pressure tensor, the heat flux and the local energy contain no intermolecular energy term. In particular, for the local energy we have the following expression:

$$\langle H\rangle = \tfrac{1}{2}m\langle u\rangle^2 + \tfrac{3}{2}kT + \Phi. \tag{9}$$

By eliminating the time derivatives of the density, the velocity, and the temperature we finally obtain

$$\partial s'/\partial t = -\partial(\langle u\rangle s')/\partial x - \partial\frac{C_x}{T}/\partial x$$

$$-\frac{P_{xy} - \delta_{xy}P}{T}\partial\langle u\rangle/\partial y - \frac{C_x}{T^2}\partial T/\partial x. \tag{10}$$

Here δ_{xy} is a Kronecker symbol and P is the average of the diagonal components of the pressure tensor:

$$P = \tfrac{1}{3}(P_{xx} + P_{yy} + P_{zz}).$$

The variation in the thermal entropy is due to three effects:

(1) a dragging effect due to the hydrodynamic motion;
(2) the heat flux which appears in accordance with the Carnot principle;
(3) sources which did not exist in the case of the statistical entropy.

When thermal equilibrium almost obtains everywhere the heat flux is low, the temperature gradient and the velocity gradient are small and the pressure is almost diagonal. The effect of the sources is negligible when compared with the effect of the entropy flux C/T. What the equations describe therefore is what in thermodynamics is called a reversible transformation. The results obtained are in accordance with the Carnot principle.

166

The arguments in this and the last section tell us nothing about the sign of the phenomena when the transformations are irreversible in nature. Nevertheless some attention must be paid to them: they suffice to show up clearly the distinction which is established between the statistical entropy and the thermal entropy. The first of these quantities has been defined with a great deal of accuracy. The second, apart from equilibrium, has been the subject of only preliminary considerations.

Intermolecular interactions must be brought into play to establish a theory of irreversible phenomena. Correlations must therefore be taken into consideration. They must be particularized by the molecular chaos hypothesis and we must finally have recourse to the Boltzmann equation. It could be concluded from this that a formula like formula (10) where no intermolecular force term appears explicitly is illusory. Such a conclusion would be inaccurate. It turns out in fact that the Boltzmann equation alters nothing in the diffusion or hydrodynamics equations, which we have manipulated, by virtue of the conservation theorems. These equations, however, do not form a complete system but are a framework which imposes certain limitations on the simple density in phase via the local averages, but is incapable of giving this density with accuracy: this rôle is reserved for the Boltzmann equation. It is the latter that allows us to show that the heat goes from hot to cold and also that the pressure tensor is constructed so as to brake the motion. The result is that the sources of the thermal entropy are positive. The study in the last section of the relations between the statistical entropy and heat was very primitive. Despite its rather simplified appearance the study in the present section, which this time concerned the thermal entropy, goes considerably deeper.

We should mention, nevertheless, that the characteristics that we ascribe here to the Boltzmann equation hold only in the case when the system is not very far from thermal equilibrium: it is the linear approximation.†

† For applications of the Boltzmann equation see the excellent treatment by Huang (1963).

References

BAKER, G. A. Jr. (1965) *Adv. Theoret. Phys.* **1**, 1.
BOLTZMANN, L. (1896) *Vorlesungen über Gastheorie*, Barth, Leipzig; translated into English by S. G. Brush as *Lectures in Gas Theory*, Univ. of California, 1964.
BRILLOUIN, L. (1922) *Ann. Phys.* **17**, 88.
CHANDRASEKHAR, S. (1960) *Principles of Stellar Dynamics*, Chicago.
DASANNACHARYA, B. A. and RAO, K. R. (1965) *Phys. Rev.* **137**, A 426.
DELCROIX, J. L. (1965) *Physics of Plasmas*, London.
DOMINICIS, C. DE (1962) *J. Math. Phys.* **3**, 983.
DYMOND, J. H., RIGBY, M. and SMITH, E. B. (1965) *J. Chem. Phys.* **42**, 2801.
FARQUHAR, I. E. (1964) *Ergodic Theory in Statistical Mechanics*, New York.
FISHER, M. E. (1964) *J. Math. Phys.* **5**, 944.
FRISCH, H. L. and LEBOWITZ, J. L. (1964) *The Equilibrium Theory of Classical Fluids*, New York.
GIBBS, J. W. (1948) *Elementary Principles in Statistical Mechanics, Collected Works*, New Haven.
GRAD, H. (1961) *Communications of Pure and Applied Mathematics*, **14**. 323.
GROOT, S. R. DE and MAZUR, P. (1962) *Non-Equilibrium Thermodynamics*, Amsterdam.
GROSS, E. (1945) *Acta Physicochimica (USSR)* **22**, 459.
GUGGENHEIM, E. A. (1945) *J. Chem. Phys.* **13**, 253.
HAAR, D. TER (1954) *Elements of Statistical Mechanics*, New York.
HAAR, D. TER (1966) *Elements of Thermostatistics*, New York.
HIRSCHFELDER, J. O., CURTISS, C. F. and BIRD, R. B. (1964) *Theory of Gases and Liquids*, New York and London.
HOLBORN and OTTO (1926) *Z. Phys.* **38**, 359.
HUANG, K. (1963) *Statistical Mechanics*, New York.
IRVING, J. H. and KIRKWOOD, J. G. (1950) *J. Chem. Phys.* **18**, 817.
KAMPEN, N. G. VAN (1955) *Physica* **21**, 949.
KINGSTON, A. E. (1965) *J. Chem. Phys.* **42**, 719.
KIRKWOOD, J. G. and BUFF, F. P. (1949) *J. Chem. Phys.* **17**, 338.
MASSIGNON (1957) *Mécanique Statistique des Fluides*, Paris.
MAYER, J. E. and MAYER, M. G. (1940) *Statistical Mechanics*, New York.
MÜNSTER, A. (1959) *Statistische Thermodynamik kondensierter Phasen, Handbuch der Physik*, XIII B.
NASA (1963) Techn. Note D. 2075 (on experiments on free fall in a tower).
NETTLETON, R. E. and GREEN, M. S. (1958) *J. Chem. Phys.* **29**, 1365.

References

ORNSTEIN, L. S. and ZERNIKE, F. (1914) *Proc. Acad. Sci. Amsterdam* **17**, 795.

PECKER, J. C. and SCHATZMAN, E. (1959) *Astrophysique Générale*, Paris.

REE, F. H. and HOOVER, W. G. (1966) *J. Chem. Phys.* **46**, 4181.

ROCARD, Y. (1932) *L'Hydrodynamique et la Théorie Cinétique des Gaz*, Paris.

ROCARD, Y. (1952) *Thermodynamique*, Paris.

ROUSSET, A. (1947) *La Diffusion de la Lumière par les Molecules Rigides*, Paris.

SALTSBURG, H. (1965) *J. Chem. Phys.* **42**, 1303.

STRATONOVICH, R. L. (1955) *J.E.T.P. USSR* **28**, 409.

SYNGE, J. L. (1957) *The Relativistic Gas*, Amsterdam.

VAN HOVE, L. (1954a) *Phys. Rev.* **93**, 268.

VAN HOVE, L. (1954b) *Phys. Rev.* **95**, 249, 1374.

VERLET, L. and LEVESQUE, D. (1967) *Physica* **36**, 244.

ZERNIKE, F. and PRINS, J. A. (1937) *Z. f. Phys.* **41**, 184.

Exercises

IN ORDER to enliven a relatively dry subject, it was decided to include a number of exercises about mixtures, although the present work does not dwell very much on the theory of such systems.

An example of mixture that is considered several times is that of a completely ionized gas. The correlations in such a medium have been covered in Appendix 8 A of Delcroix's book (1965).

CHAPTER 1

§ 1.4. (a) Generalize formula (4) to evaluate

$$\langle \varDelta^3 \rangle \quad \text{and} \quad \langle \varDelta^4 \rangle.$$

Attention must be paid to the numerical coefficients. It is advisable to give the results in the most symmetrical form possible.

Express the inequality $\langle (\varDelta^2 - \langle \varDelta^2 \rangle)^2 \rangle \geqq 0$ in terms of the density.

(b) Generalize formula (4) to the case of a mixture of two kinds of particles, a and b. In other words, calculate the variances

$$\langle \varDelta_a^2 \rangle, \qquad \langle \varDelta_b^2 \rangle.$$

Calculate also the average

$$\langle \varDelta_a \varDelta_b \rangle.$$

The results will be reconsidered in a subsequent exercise.

§ 1.12. What does the equation of energy (37) become if the kinetic energy $\frac{1}{2}m \langle u \rangle^2$ is subtracted from H?

Exercises

CHAPTER 2

§ 2.2. (a) Give formula (4) explicitly for the following cases:

$$M = 1, \quad M = 2, \quad M = 3.$$

Do the same for formula (5). Show that, term by term, formula (5) is a consequence of formula (4).
  Prove formula (6).

  (b) Give the general form of equation (4) of § 1.4,

$$\langle \varDelta^2 \rangle = \int_A n_1 \, d^3 r_1 + \int_A \int_A (n_{12} - n_1 n_2) \, d^3 r_1 \, d^3 r_2$$

for the case when the occupation is ill defined.
  Use the hypothesis of molecular disorder to explain why the expression

$$\langle \varDelta^2 \rangle = \int_A n_1 \, d^3 r_1 + \int_A \int_V (n_{12} - n_1 n_2) \, d^3 r_1 \, d^3 r_2$$

is a good approximation of the former expression when the volume A is not too small. Show that, on the other hand, this approximation is meaningless when the occupation is well defined.

  (c) The results of exercise 1.4 (a) can also be generalized. Apply them to the case of a completely ionized gas made up of (positive) protons and (negative) electrons which has the following characteristics: (i) the gas is neutral; (ii) it is homogeneous throughout volume A and its neighbourhood; and (iii) the gas is in thermal equilibrium and the correlations depend only on the distance.

  These correlations are given by the following approximate formulae:

$$\varepsilon_{12++} = \varepsilon_{12--} = -\frac{q^2}{kT} \frac{1}{r} e^{-r/h},$$

$$\varepsilon_{12+-} = \varepsilon_{12-+} = -\varepsilon_{12++},$$

q being the charge of the proton and h the Debye length:

$$h = \sqrt{\frac{kT}{8\pi n_+ q^2}}.$$

Then show that

$$\langle \varDelta_+ \rangle^2 \quad \text{and} \quad \langle \varDelta_- \rangle^2$$

have a value which is half of that corresponding to the perfect gas.
  Show that the variance of the electric charge contained in A,

$$q^2 \langle (\varDelta_+ - \varDelta_-)^2 \rangle,$$

vanishes.

172

§ 2.6. (a) Prove formula (5) by expanding the first terms of exp G.

(b) Using the hypothesis of molecular disorder, show that in formula (12) the domains of integration only affect the immediate neighbourhood of the point x_1, y_1, z_1.

What does formula (13) become when we imagine x_2, y_2, z_2 to be far from x_1, y_1, z_1?

§ 2.7. Prove, in detail, formula (6).

§ 2.10. Prove, in detail, formula (8).

§ 2.11. Prove, in detail, formulae (11) and (12).

CHAPTER 4

Identical point molecules of mass m are held near the origin of coordinates by a force field, of which the potential energy for the particle J is

$$V_J = \tfrac{1}{2} m\Omega^2 (x_J^2 + y_J^2 + z_J^2),$$

where Ω is a constant frequency.

On the other hand, there exists an intermolecular energy W_{JK}, which is a function of distance, and which, although very weak, is essential for the thermalization and can be neglected, except in the third question.

(a) The system contains exactly N particles. It is at temperature T. Show that its average energy is

$$U = 3NKT,$$

and show that

$$R = \sqrt{\langle x_J^2 + y_J^2 + z_J^2 \rangle} = \left[\frac{3kT}{m\Omega^2} \right]^{1/2}.$$

(b) The occupation is ill defined. Prove the following expression for the chemical potential:

$$\omega = kT \ln \langle N \rangle \left(\frac{h\Omega}{2\pi kT} \right)^3,$$

and the following expression for the entropy:

$$S = k \langle N \rangle \left\{ 4 - \ln \left[\left(\frac{h}{2\pi kT} \right)^3 \langle N \rangle \right] \right\}.$$

(c) The intermolecular energy is equal to

$$-w_0 (w_0 > 0) \quad \text{for} \quad r_{JK} < \sigma,$$
$$0 \quad \text{for} \quad r_{JK} > \sigma.$$

173

Exercises

Estimate, very approximately, the correction to the internal energy due to the presence of the intermolecular interaction, using the following values of the parameters: $T = 300°\text{K}$, $m\Omega^2 = k \cdot 300$ erg cm^{-2}, $\sigma = 10^{-7}$ cm, $\langle N \rangle = 10^{20}$, $W_0 = 10k$ erg, $k = 1\cdot 4 \cdot 10^{-16}$ erg $(°\text{K})^{-1}$.

Do the same for the fluctuations or, more precisely, for $\langle \Delta N^2 \rangle$ (this really applies to later chapters).

$$\left[\int_{-\infty}^{+\infty} e^{-ax^2}\, dx = \sqrt{\frac{\pi}{a}}\,, \qquad \int_{-\infty}^{+\infty} x^2 e^{-ax^2}\, dx = \frac{1}{2}\sqrt{\frac{\pi}{a^3}} \right].$$

CHAPTER 5

§ 5.4. (a) Formula (7) deals with infinitesimal variations when the temperature and chemical potential are fixed. What relation can be obtained if the condition $\delta\omega = 0$ is replaced by the condition $\delta \langle N \rangle = 0$?

(b) Generalize formula (7) to the case of a mixture. Apply the results to the case of the ionized medium defined in exercise 2.2 (c). Estimate particularly the infinitesimal variation of the electric density in a region where the unperturbed medium is neutral and homogeneous. Deal with the case where the perturbation is caused by a fixed point charge, brought in from the outside, and show that the perturbation of the electric density is then zero.

§ 5.6. (a) In a fluid which is homogeneous in first approximation one studies the perturbation caused by gravity. Show that formulae (7) of § 5.4 and (8) of § 5.6 lead to the same results.

(b) Some freely moving electrons are contained in a sphere of radius R. They are in thermal equilibrium at temperature T.

Their distribution inside the sphere is studied under different conditions. To do that, the first recurrence equation

$$kT\frac{\partial n_1}{\partial x_1} = n_1 X_1 + \int n_{12} X_{12}\, d^3 r_2$$

is used with

$$X_{12} = -\frac{\partial W_{12}}{\partial x_1}\,, \qquad W_{12} = \frac{q^2}{r_{12}}\,.$$

(i) The sphere only contains electrons, and the self-consistent field approximation is sufficient. Derive an expression for the self-consistent field and give it as a function of $n(r)$ and of the radius.

(ii) $n(r)$ is written in the form

$$n(r) = n(0)\,(1 + f(r)).$$

If the temperature is raised (for fixed $n(0)$), show that, approximately,

$$f(r) \sim \frac{1}{6} \frac{r^2}{h^2}$$

with

$$h = \left[\frac{kT}{8\pi n(0) q^2} \right]^{1/2}.$$

For $R = 1$ cm, $n(0) = 10^6$ cm^{-3}, calculate T, if $f(R) = 0\cdot1$.

Calculate h. [$k = 1\cdot38 \cdot 10^{-16}$ erg ($^\circ$K)$^{-1}$, $q = 4\cdot8 \cdot 10^{-10}$ e.s.u.]

(iii) The sphere contains positive fixed charges (a somewhat artificial situation!), uniformly distributed, with density

$$n(0) (1 - a) |q|,$$

a being a small parameter. Calculate $f(r)$ in the same way as above.
The temperature being the same as the one calculated in (ii), but the density now being 10^{18} cm^{-3}, calculate a if $f(R)$ must still be equal to $0\cdot1$. Find h.

(iv) For a better justification of the approximations which have led to the results of (iii), the effect of correlations between electrons are estimated approximately. In this case they are given by the formula

$$a_{12} = 1 - \frac{q^2}{kT} \frac{1}{r_{12}} e^{-r_{12}/h}.$$

Show that the effect of these correlations will be to diminish slightly $f(R)$ to an extent that can be determined numerically.

CHAPTER 6

§ **6.1.** Derive (in the simplest way) the approximation (2) from the recurrence equations of the second kind.

§ **6.3.** (a) Continuing from the preceding exercise, derive the following approximations of equations (5),

$$C - \beta V_1 = \ln n_1 - \int g_{12} n_2 \, d^3 r_2,$$

from the recurrence equations of the second kind. The parameter C cannot be determined by this method.

Exercises

(b) Suppose that the particles, instead of occupying arbitrary positions in space, are fixed at lattice sites. Physically, it could be a crystal lattice, but the kind of lattice is not essential for what follows.

Equation (5) can be used to deal with this case, provided that the integrations are replaced by summations over a discrete argument.

Consider the special case where there is no interaction between the particles, with the proviso that no two particles can ever occupy the same lattice site. Show that then $\int g_{12}n_2 d^3r_2$ must be replaced by n_1; $\frac{1}{2}\int g_{12}g_{13}g_{23}n_2n_3 d^3r_2 d^3r_3$ must be replaced by $\frac{1}{2}n_1^2$; and the following term must be replaced by $\frac{1}{3}n_1^3$.

One obtains then a condensed version of equation (5) by extrapolation. Compare the result with the basic (quantum mechanical) formula of Fermi statistics.

§ 6.7. (a) In the general equation for the pressure

$$P = nkT - \frac{2\pi}{3} \int_0^\infty n_{12}(r) \frac{dW}{dr} r^3 \, dr$$

substitute the following approximation for n_{12}:

$$n_{12} = n^2 \exp(-W/kT).$$

Hence deduce expression (3) of § 6.7 for the second virial coefficient.

N.B.: There exist formal examples in which $W(r)$ is a discontinuous function of distance—the most simple of these is the example of hard spheres—and when W is a discontinuous function, equation (1) has not got a very clear meaning.

Is this a serious difficulty?

(b) A uniform gas is moderately dense. The potential energy $W(r)$ is made up of:

a repulsive core, which is infinite for $r < \sigma$, and
a constant attractive part, W_0, for $\sigma < r < 2\sigma$,
    with W_0 small compared to kT.

Consider the first corrections to the theory of a perfect gas, namely binary effects.

Express the pressure, the internal energy, the entropy, and the specific heat at constant density as functions of density and temperature.

(c) (This is a tricky exercise, because it presents in classical theory a question which relates partly to quantum theory. However, it prepares the reader for the study of Saha's equation, which controls the phenomena of dissociation or ionization in a gas.)

A gas in thermodynamic equilibrium is made up of two types of particles—which are force centres. All interaction between similar particles will be neglected, but, on the other hand, there is a basically attractive short-range force between any two particles of different types. There is no applied field.

176

The following symbols are used:

a and b: subscripts corresponding to types of particles,
n_a: density (number of particles per cm^3),
r_a: position,
p_a: momentum,
m_a mass,
f_a: normalized Maxwell distribution,
$W = W(|r_a - r_b|)$: potential energy for an a, b, pair,
$\mu = \mu(r_a, p_a; r_b, p_b)$: reduced mixed density in phase.

We define

$$\mu \, d^3r_a \, d^3p_a \, d^3r_b \, d^3p_b$$

as the probability of finding at the same time particle a in $d^3r_a d^3p_a$ and particle b in $d^3r_b d^3p_b$.

f_a is given by

$$f_a = C_a \exp(-\beta p_a^2/2m_a)$$

with

$$C_a = (2\pi m_a kT)^{-3/2}. \tag{1}$$

It can be assumed—by analogy with what occurs for particles of the same type—that the density in phase μ can be expressed in the following way:

$$\mu = n_a n_b f_a f_b \exp(-\beta W). \tag{2}$$

The medium is uniform.

(1) The potential energy is defined as follows:

$$W = \text{constant} = -k\theta \quad \text{for} \quad |r_b - r_a| < \sigma \quad \text{with} \quad \theta > 0,$$
$$W = 0 \qquad\qquad\qquad \text{for} \quad |r_b - r_a| > \sigma.$$

Evaluate the average number n per cm^3 of b particles which are present within the radius of action of an a particle.

(2) Numerical application: $n_a = n_b$ is the normal density; $\sigma = 10^{-8}$ cm. Calculate n/n_a for the cases of $T = 4\theta$ and $T = \frac{1}{4}\theta$.

(3) Considering a couple of particles of opposite types, the motion of the centre of mass can be separated from the relative motion by introducing the new co-ordinates:

$$r = (m_a r_a + m_b r_b)/m, \qquad R = r_b - r_a$$

and the new momenta

$$p = p_a + p_b, \qquad P = M\left(\frac{p_b}{m_b} - \frac{p_a}{m_a}\right),$$

$$m = m_a + m_b, \qquad M = m_a m_b/(m_a + m_b).$$

Express the probability of having simultaneously:

$$r \text{ in } d^3r, \quad R \text{ in } d^3R, \quad p \text{ in } d^3p \quad \text{and} \quad P \text{ in } d^3P$$

Exercises

and then the simpler probability of having simultaneously:

r in $\mathrm{d}^3 r$, R in $\mathrm{d}^3 R$, P in $\mathrm{d}^3 P$ and p unspecified.

(4) With the help of the preceding result and the same law of force as in (1) give the expression for the average number n', per cm^3, of b particles present in the radius of action of an a particle, assuming that the internal energy of the ab couple is negative.

(5) Calculate n'/n_a with the assumptions given under (2), but only for the case $T = 4\theta$; assume $m_a \ll m_b$.

(6) The situation foreseen in (4) could be considered as a definition of the formation of a molecule. The result obtained would therefore be the Saha formula. Underline the weak points of such a deduction without overlooking the satisfactory ones.

CHAPTER 9

§ 9.2. (a) When a low deusity gas is in thermodynamic equilibrium its properties are defined by the laws of a perfect gas. The molecular interactions do not enter in any way.

It is not the same in non-equilibrium conditions, which are described by Boltzmann's equation. The aim of the following exercises is to discover how far one can go in the study of non-equilibrium situations without introducing intermolecular forces.

The simplest way of working this out is to replace the equation (2) of § 9.2 by the following equation, clearly easier for calculations (the index 1 is omitted):

$$\left(\frac{\partial}{\partial t} + \frac{p}{m} \frac{\partial}{\partial x} + X \frac{\partial}{\partial p} \right) \mu = 0. \tag{1}$$

(We restrict ourselves to the case where there is no applied field: the medium is undefined.) What is the trajectory of the particles? Express $\mu(r, p, t)$ as a function of $\mu(r, p, 0)$. Under what conditions does the medium, thermalized initially, remain thermalized? A distinction is to be made between the thermalization in laboratory conditions and thermalization under uniform motion conditions.

A solution of the form $\mu = \mu_0 + \mu'$ is studied, where μ_0 is the density in a simple thermalized phase and μ' is a perturbation. We are trying to represent μ' as the superposition of complex plane waves.

$$F(p) \exp i[\omega t - (s \cdot r)]. \tag{2}$$

Show that these waves are singular. Can one speak of the speed of sound?

(b) The question is tackled in an entirely different way. It is assumed that at each instant the phase density is of the form:

$$\mu = n(2\pi mkT)^{-3/2} \exp\left[-(p - p_0)^2/2mkT\right], \tag{3}$$

n, p_0, T being functions of space (more simply only of x; and let q_0 and r_0 be zero).

Give the expressions at each point of the average local speed, the pressure, energy density and heat flux.

Introduce these quantities in the diffusion equations (Chapter 1: hydrodynamic equations). Then check that an infinitesimal perturbation of the thermodynamic equilibrium is propagated with the speed of sound.

We have therefore succeeded, by cheating—that is, by only keeping in equation (1) some of its results—in obtaining a theory suitable for a perfect gas. This is because the molecular reactions which were incorrectly neglected maintain quasi-Maxwellian conditions. This theory naturally has its limits: irreversible phenomena are still outside its scope. This is an essential feature when we attack the difficult relativistic problems (Synge, 1957). However, version (a) of the problem is of some use: it is applicable in the case of very rarified media.

(c) Attempts can be made to improve the theory without becoming involved in enormous difficulties. We restrict ourselves to cases where the forces, still short-range ones, are quite weak, so that we can use the collective field approximation. Equation (1) is therefore replaced by Vlasov's equation (8) of § 2.7:

$$\left(\frac{\partial}{\partial t} + \frac{p}{m} \frac{\partial}{\partial x} + X'' \frac{\partial}{\partial p}\right)\mu = 0. \tag{4}$$

Show that the statistical entropy, reduced to its first term,

$$S = -k \int \mu(1 - \ln \Gamma \mu)\, d\Omega, \tag{5}$$

is conserved.

We assume that the molecular density varies slowly from point to point. Show that the collective field is approximately given by

$$X'' = -A \frac{\partial n}{\partial x}, \tag{6}$$

the coefficient A being positive in the case where the force is essentially repulsive.

Study now solutions of the form

$$\mu = \mu_0 + \mu'$$

with

$$\mu_0 = n_0(2\pi mkT)^{-3/2} \exp\left[-p^2/2mkT\right], \tag{7}$$

μ' being a slight perturbation. Justify the approximate equation

$$\left(\frac{\partial}{\partial t} + \frac{p}{m} \frac{\partial}{\partial x}\right)\mu' = A \frac{\partial n'}{\partial x} \frac{\partial \mu_0}{\partial p}$$

179

Exercises

where
$$n' = \int \mu' \, \mathrm{d}^3 p.$$

Show that the entropy (5) remains conserved to first order. Derive the representation of μ' in the form of plane waves of type (2). Mathematical distribution theory (van Kampen, 1955) must be used. Show that in this approximation, as in exercise (a), it is impossible to find the speed of sound.

(d) While keeping the approximation of the collective field and formula (6), use the cheater's method and calculate the speed of sound, following the instructions of exercise (b). Calculate finally the entropy.

§ 9.3. Show that entropy (2) has an upper limit. In order to do this the inequality
$$-\mu \ln \Gamma\mu + \mu \ln \Gamma\mu_0 \leqq \mu_0 - \mu$$

is used, where μ_0 will be a conveniently chosen density in phase—for example,
$$\mu_0 = C \exp\left[-\beta_0 p^2 / 2m - \beta_0 V\right].$$

Index

Index

Reduced density 37, 71–7
Reduced quantities xii
Regressive procedure xii
Resistivity 57

Saha formula 176, 178
Second virial coefficient 114, 176
Self-consistent field 11, 29, 45
Self-consistent potential 11
Short-range forces 8
Simple density 7, 24
Sound velocity 178–80
Specific heat 144
State of rest xii
Statistical entropy xiii, 58–77, 152–4,
    160–5
Statistically perfect fluid 6
Stellar dynamics xvi
Stirling's formula 45
Stochastic theory xv, xvi
Superheated liquid 125
Supersaturated vapour 125
Surface tension 128, 132–3

Thermal entropy 164–5
Thermal equilibrium 85–124, 165
Thermal waves 134–51
Thermalization 92–5, 164
Thermodynamic entropy 60, 66, 164
Thermodynamic equilibrium xii,
    xiii, 12, 78–84, 159, 164
Turbulence 34

Uniform medium 96, 104–11, 132,
    143
Ursell's equation of state 107, 109,
    113

Van der Waals equation 116, 125,
    126, 128, 131
Van der Waals forces 113, 115, 123
Variance 4
Velocity distribution 93
Virial coefficient 113–18
Vlasov equation 46, 179